Noble Cows and Hybrid Zebras

Harriet Ritvo

NOBLE COWS
— and —
HYBRID ZEBRAS

Essays on Animals and History

University of Virginia Press   *Charlottesville and London*

University of Virginia Press
© 2010 by Harriet Ritvo
All rights reserved
Printed in the United States of America on acid-free paper

First published 2010

1 3 5 7 9 8 6 4 2

Library of Congress Cataloging-in-Publication Data
Ritvo, Harriet
Noble cows and hybrid zebras : essays on animals and history / Harriet Ritvo.
    p.   cm.
Includes bibliographical references and index.
ISBN 978-0-8139-3060-2 (cloth : alk. paper)
1. Human-animal relationships—History. 2. Animal welfare—Public opinion—History. 3. Animals and history. 4. Animals (Philosophy)—History. I. Title.
QL85.R58 2010
304.2′709—dc22

2010020377

Illustration credits follow the index.

In memoriam
Zelma Ritvo
1918–2010

> The innumerable past and present inhabitants of the world are connected together by the most singular and complex affinities.
>
> CHARLES DARWIN, *The Variation of Animals and Plants under Domestication*

Contents

	Introduction	1
1.	Sex and the Single Animal	13
2.	Learning from Animals: Natural History for Children in the Eighteenth and Nineteenth Centuries	29
3.	Toward a More Peaceable Kingdom	50
4.	Animal Consciousness: Some Historical Perspective	63
5.	Plus Ça Change: Antivivisection Then and Now	73
6.	Mad Cow Mysteries	91
7.	Understanding Audiences and Misunderstanding Audiences: Some Publics for Science	103
8.	Foreword to Charles Darwin, *The Variation of Animals and Plants under Domestication*	123

9. Race, Breed, and Myths of Origin:
   Chillingham Cattle as Ancient Britons   132

10. Possessing Mother Nature: Genetic Capital
    in Eighteenth-Century Britain   157

11. Our Animal Cousins   177

12. Counting Sheep in the English Lake District:
    Rare Breeds, Local Knowledge, and
    Environmental History   186

13. Beasts in the Jungle (or Wherever)   203

    Bibliography   213

    Index   227

Noble Cows and Hybrid Zebras

Introduction

In the early 1980s, when I started working on *The Animal Estate: The English and Other Creatures in the Victorian Age*, it was considered both unusual and eccentric (which are not at all the same thing) to take animals seriously as historical subjects. Even the fact that Keith Thomas—a supremely respectable historian of early modern Britain, who had previously written about the supremely respectable topic of religion—dealt with animals as well as vegetation and landscape in his *Man and the Natural World,* which was published in 1983, did not redeem the subject in everyone's eyes. My work was once introduced as "the weirdest" of the "many weird things that have been coming out of the humanities lately." Things have changed greatly in the intervening years, not only in the discipline of history but throughout the humanities and social sciences. Scholars now lavish attention on nonhuman animals. What has been called "the animal turn" basks in the full panoply of academic institutionalization: conferences, journals, book series, websites, societies, and even a few majors in "animal studies."

Of course there is a sense in which animals have always been a part of humanistic studies—but most often they have been simultaneously mainstreamed and submerged. For example, animals have often figured in historical accounts, but they have seldom been their focus. Prehistoric animals provided food and fetishes for early hunters. The development of agriculture depended in part on the domestication of ancient ungulates (although the

---

This introduction originally appeared as "Animal Planet" in *Environmental History* 9, no. 2 (2004): 204–22, and is reprinted by permission of Oxford University Press.

domestication of ancient plants normally gets top billing) and, with regard to later periods, livestock has traditionally attracted the attention of economic historians who focus on agriculture. Wild animals have often represented nature (however nature has been understood) in religious and scientific thought. Domesticated animals have taken their place alongside oppressed human groups as the objects of outrage and amelioration. Primates and platypuses, among many others, have figured prominently in the development of modern biology. Some distinguished individual animals, from Jumbo to Flush to Seabiscuit, have even become the heroes of their own lives, and some of their owners, minders, and pursuers have achieved similar distinction.

So it is not so much that the absolute range of subjects available to humanists has increased recently, but that the understanding of how animals figure in them has changed. It is difficult to specify exactly in what this change consists, especially since understandings vary greatly among interpreters, and the category "animal" is far from monolithic. But it might be fair to say that the interests or at least the perspectives of animals are now more likely to be taken into account, along with those of people. Even in fields where animal topics have been routine, such as agricultural history, the farmyard creatures are less frequently abstracted through quantification, and more frequently presented as individuals, or at least groups of individuals. As has been the case with some other areas of scholarly investigation, academic interest tends to resonate with issues in the news. Animals can be seen as the latest beneficiaries of a protracted movement toward greater inclusiveness. Just as organized labor, the civil rights movement, decolonization, and the women's movement inspired sympathetic researchers, so have, in their turn, the advocates of hunted whales, poached tigers, abandoned dogs, and overcrowded pigs.

Like scientists, humanists and social scientists have become increasingly concerned with the relationships among animal species, including our own, with the ethical and political questions raised by our treatment of members of other species, and especially with whether the answers to these perennial questions have been altered, or should be, by recent work in genomics, bioengineering, and allied pursuits. The resulting investigations have revealed many conflicting commitments, based on politics, disciplinary affiliation, and relation (or absence of relation) to commercialization and industry, as well as on underlying philosophical positions. Of course, it is a commonplace that the study of animals tends also to function in this way. As Claude Lévi-Strauss put it long ago, "Animals are good to think [with]."[1]

Introduction

Nevertheless, there are important convergences with regard to a couple of the basic questions (or maybe, a single question asked in a couple of different ways): what is an animal? and are we them? Most scholars who specialize in the study of animals believe that human beings fall within that category. This is as true of scientists, who locate *Homo sapiens* within the primate order, along with lemurs, monkeys, and other apes, as it is of humanists (whether or not they are posthumanists—a self-description used by some members of the animal studies community) who claim kindred in footnotes or parentheses. (Here is my own declaration: I share the view that people are animals.) This consensus is heartening, but not entirely persuasive. What seems most interesting about this chorus of response is not its unison, but its tone—and the fact that the existence of the consensus is worthy of note. However clearly and confidently they are made, such assertions often seem defensive or even strident. Indeed, the compulsion to make them strongly signals a persistent context of semantic and cultural tension: Why do we have to keep reiterating this point again and again, over the decades and even the centuries? The consensus of experts has only intensified over time, so what continues to be the problem?

This tension is evident even on the level of practice, where it might be supposed to have the least impact. The persistent reluctance of many taxonomists to locate ourselves and our closest extinct relatives in the family Pongidae, which usually includes bonobos, chimpanzees, gorillas, and orangutans, rather than in the more exclusive family Hominidae, reserved for australopithecines and humans, is one response to this uneasiness. The labels on the ape exhibits at many zoos show equal restraint, perhaps under greater pressure—although the boldest among them gesture at human proximity by placing mirrors next to the cages. And abstract definitions also embody the uncomfortable conjunction of likeness and otherness. For example, the *Oxford English Dictionary* includes two distinct senses in its entry for *animal*. The first sense, illustrated with learned examples ranging from John de Trevisa to Thomas Henry Huxley, includes all living things that are not plants; the second sense, illustrated mostly with literary quotations, is less inclusive and more popular: "in common usage: one of the lower animals; a brute, or beast, as distinguished from man." Thus the attempt to classify animals ends up also implicating human categories—drawing a distinction between the learned and the less learned.

One possible explanation for the disjunction between the expert view

Captive orangutan dressed to look human, from Charles Knight, *Pictorial Museum of Animated Nature*, 1844.

and the popular sense is the tendency of learned experts to see their work in isolation—that is, as unconstrained by their own context—no matter how contextually nuanced their theoretical orientation or methodological approach may be. But despite their careful definitions and their forceful assertions, scholars are inevitably influenced at least as much by the common usage of the terms that they deploy, as they are by their more rarefied and specialized senses. Like Archimedes, whose irremediable terrestriality prevented him from moving the earth, humanists (and posthumanists—although that term embodies a lot of wishful thinking) cannot escape their real-world location. Ignoring the conventional meanings of words risks the fate of Humpty Dumpty, who claims in *Through the Looking Glass* that when he uses the word *glory,* it means "a nice knockdown argument," and further that, with regard to meaning, "the question is . . . which is to be master"—himself or the word.[2]

Introduction

Humpty Dumpty was confident about the answer—although he discovered that it is unwise to assume that vocabularies can be so easily dominated. With regard to the study of animals, the unruly multiplicity of senses often means that overt claims of unity (humans are animals) unintentionally work to reinforce the human-animal boundary that they have been formulated to dissolve. Such claims incorporate a grudging acknowledgment that this boundary is widely recognized and powerfully influential. Why else would it be continually necessary to deny its validity—or to remind ourselves of its arbitrariness? Further, like clichéd metaphors that turn out to be only half-dead, they may bring buried assumptions and understandings into the full light of consciousness, thus paradoxically highlighting contradictions that might otherwise not have been articulated. Or, to put it another way, chickens should not be counted before they hatch.

The traditional boundary between humans and animals has nevertheless become increasingly permeable, at least from the perspective of some inhabitants of the more affluent parts of the world, as the result of a gradual and protracted shift in attitudes. Although, for example, most Victorians explicitly acknowledged the robustness of that boundary, they also encountered some longstanding challenges to this understanding of the relationship between people and other creatures. The most formidable of these was evolution, or, to put it in the terms of the eighteenth century, the notion that the chain of being represented relationships that were dynamic rather than static. Evolution was not a new idea in the nineteenth century—for example, Charles Darwin's grandfather Erasmus had written eloquently about it—but it did problematize intraprimate resemblances that were already obvious. From its Enlightenment beginnings, most formal taxonomy recognized not only the many general correspondences between people and what were then known as quadrupeds (rechristened *mammals* by Linnaeus) but also the particular similarities that human beings shared with apes and monkeys. It was the nonfunctional details that proved most compelling (that is, the hardest to explain away): the shape of the external ear, for example, or the flatness of fingernails and toenails. On this basis, not only did Linnaeus locate people firmly within the animal kingdom, but he constructed the primate order to accommodate humans, apes, monkeys, prosimians, and bats.[3] Indeed, in some versions of his system *Homo sapiens* was not the only living occupant of the genus *Homo*, as he (or she) is today. This categorization was far from universally persuasive; on the contrary, it provoked a great deal of resistance, even among learned experts, some of whom proposed alternative taxonomies that separated people more dramatically from other creatures, or omitted people altogether.

In order to minimize such reactions, Charles Darwin tried to explain his theory of evolution by natural selection in the least alarming terms possible, assuring fellow naturalists, for example, that they would not have to change any of their taxonomic categories as a result of his arguments. On the contrary, he explained, he was simply providing a better explanation for their well-established practices. Perhaps his reassurances were too persuasive. In any case, after the publication of *On the Origin of Species* in 1859, as well as before, for most educated Victorians the default answer to the question, "Are people animals?" would still have been "No." As Darwin sadly noted at the end of *The Descent of Man,* written a decade after the appearance of the *Origin,* "The main conclusion arrived at in this work, namely that man is descended from some lowly-organised form, will, I regret to think, be highly distasteful to many persons."[4] Even Alfred Russel Wallace, who famously converged with Darwin on natural selection, ultimately differed with him on this point: he felt that even if the human body was the material product of natural selection, the human mind and spirit had a different, nonevolutionary origin. In the intervening century and a half, Darwin's (and Wallace's) evolutionary theory has been enshrined as biological orthodoxy, but Darwin's observation is far from obsolete.

But in the nineteenth century as now, the explicit denial of kinship between ourselves and the rest of the zoological world would not have been the end of the story. As is still the case, cognitive dissonance seems to have been among the least troublesome of mental conditions. Alongside formal assertions of extreme difference—that animals' lack of a soul or of elevated mental abilities constituted an insurmountable barrier separating them from people—existed many informal acknowledgments of similarity or even identity. For example, when breeders castigated the lasciviousness of their female animals, or bemoaned their reluctance to accept the mates selected for them, they channeled the outrage of the flouted Victorian paterfamilias, faced with a daughter who stubbornly insisted on charting her own romantic course—thus conflating the willful maiden with the intransigent mare or bitch. When they celebrated the purity of pedigreed animals they confirmed and exalted the prestige of their own ancestry. Both lines of descent were memorialized in volumes referred to as "stud books," and outstanding specimens of both (that is, their portraits in oils) graced the walls of many distinguished residences.

Introduction

Good animal behavior, especially the loyal devotion of dogs and horses, was described in terms equally applicable to human servants or employees. The defenders of abused animals borrowed the rhetoric of antislavery activists and advocates of factory reform. Sometimes such resonances were figured as metonymy, emphasizing similarity, and sometimes they were figured as metaphor, emphasizing difference. But whether the animal analog was wild or domesticated, primate or ungulate or carnivore, continuity and discontinuity were inextricably intertwined.

The admirable behavior of pets or livestock animals could even be appreciatively characterized as "sagacity" or, more rarely, "intelligence," perhaps because their thoughts and actions nevertheless seemed so different from those of their admiring and affectionate proprietors. The intellectual powers of the animals anatomically closest to humans inspired more complex responses, reflecting anxiety as well as appreciation. But the appeal of proximity apparently trumped its disconcerting aspect—or at least the conventions for displaying live primates unambiguously emphasized resemblance. Zoo apes and sideshow monkeys were clothed in jackets and dresses; they ate with utensils and drank from cups; they appeared to enjoy smoking cigarettes and leafing through illustrated books. The guardians of public morality kept a watchful eye on all animal attractions (a habit dating back to the Puritan disapproval of animal combat), worried that they were potential sites of unedifying behavior on the part of both exhibited creatures (so that the feeding of live prey to carnivores was prohibited) and raucous human observers (so that admitting the lower classes into zoos was initially controversial). They did not, however, seem to be particularly sensitive to the blasphemous undertones of such presentations.

The pages of many natural history books and travel accounts contained still more suggestive evidence of closeness: suggestions, speculative but compelling, that humans might be the objects or even the originators of potentially fruitful relationships with orangutans and chimpanzees, although scientific accounts of such episodes tended to be carefully distanced by skepticism or censure.[5] Nevertheless, it was respectably and repeatedly reported that orangutans "have been known to carry off negro-boys, girls and even women . . . as objects of brutal passion" and that "lascivious male apes attack women," who "perish miserably in the brutal embraces of their ravishers."[6] Edward Tyson had implicitly acknowledged such reports when he assured readers of his anatomy of a chimpanzee that "notwithstanding our *Pygmie* does so much resemble a *Man* . . . yet by no means do I look upon it as the Product of a *mixt* generation."[7] No offspring of such a conjunction was ever reliably reported, but outside the community of experts, claims could be less restrained. There do not seem to have been any Victorian successors to the "poor miserable fellow" who, according to a mid-seventeenth-century report, was tempted into bestiality with a monkey, "not out of any evil intention . . . , but only to procreat a Monster, with which . . . he might win his bread."[8] In sideshows, however, non-Europeans who were unusually hairy or adept with their toes were repeatedly ballyhooed as products of an ape-human cross, or, after *On the Origin of Species* was published, as missing links. Of course, *mutatis mutandis,* modern biotechnology has raised many of the same issues—and with regard to a much wider range of possible biological partners.

Domesticated animals, both livestock and pets, were omnipresent in Victorian daily life, and in the thoughts and feelings of the people who lived it. So close were many relationships that criticizing accounts of them as anthropomorphic can seem beside the point. Then as now, some pets, for example, really did belong to human families in all but the narrowest biological sense. And at the other end of the affective scale, relationships between some working animals and their owners strongly resembled relationships between some human laborers and their employers. (This is the well known point of Anna Sewell's *Black Beauty,* for example.) As with exclusively human connections, neither familial relationships nor professional ones necessarily guaranteed good treatment, but they did guarantee at least some degree of intimacy. The notion of anthropomorphism eliminates the possibility of easy interspecific slippage, and erects or resurrects a barrier that may not have been perceived by any of the individuals involved, human or otherwise. Indeed, the very term *anthropomorphism* is problematic, since it implicitly disparages the possibility that humans and nonhumans share perceptions, behaviors, and responses. Thus it ineluctably privileges the binary distinction between humans and other animals.

But if exploring this history sheds a unique light on human experience in the Victorian era and before, it also emphasizes the extent to which the experiences of humans and at least some other animals were interdigitated at that time and place. The likeliest targets of unconscious identification and projection were the animals who were most like people, either because they looked like people or because they were members (whether under- or hyper-privileged) of the same society. Animals outside these overlapping circles of familiarity were much less likely potential surrogates. Even accessible wild animals, whether roaming free in the woods or confined in zoos and menageries, required an additional layer of figuration, and this was much more likely to be conscious. The African elephant Jumbo, for example, could not be called a natural representative of the British nation, although he turned out to be a compelling one, at least for a brief period.

Animals less available for incorporation into the human sphere—the inhabitants of remote regions, as well as animals without fur or (more extreme) without a backbone—also attracted a great deal of interest. Then as now, they were the subjects of scientific study and amateur fascination, which resulted in numerous books and massive collections. But with a few exceptions the interest was of a different kind. For example, the social insects (ants and bees) whose economic organizations seemed to replicate those of people could figure as models for human behavior—as in the hymn that begins "How doth the little busy bee / Improve each shining hour."[9] Or, in a more attenuated metaphor, aquatic creatures could, in the spirit of "ontogeny recapitulates phylogeny," be seen to figure in prenatal human development as well as in remote human ancestry. Thus in the phantasmagoria toward the end of Charles Kingsley's novel *Alton Locke,* the hero imagines, "I was at the lowest point of created life; a madrepore rooted to the rock; . . . I was . . . a crowd of innumerable polypi; . . . I was a soft crab . . . ; I was a remora" and so on.[10]

But on the whole, such creatures seemed so alien that the use of the blanket term *animal* to cover them all brings the term itself into question. This expansive and promiscuous usage epitomizes a serious difficulty implicit in the abrogation of the dichotomy between humans and other animals: the elimination of one boundary seems to require the establishment of another or others, although the location of replacement boundaries is equally problematic. If no obvious gap can be discerned between most kinds of animals and those kinds most similar to them, then large gaps emerge when very dissimilar animals are juxtaposed. The claim that people are like cats or beavers or hippopotami (that they belong in the same general category with those kinds of creatures) is not the same as the claim that they are like jellyfish or fleas or worms. Both claims are interesting, and both are compelling, but they make sense in different contexts.

That is, they make sense in different human contexts. The nonhuman perspective is ultimately a matter for speculation, but it seems likely that the living world is differently organized in the view of cats or beavers or hippopotami. To begin with, if they understand the world in binary terms, it is probably not according to the human/nonhuman binary that seems natural to us (or at least, history suggests, to many people). Certainly, the experience of the gorilla Koko, a longtime resident of Palo Alto, California, who has shown romantic interest in male humans and has experienced the pleasures of pet ownership, suggests an alternative taxonomy, as does the behavior of many domestic dogs.[11]

Confusion about the appropriate context—or intentional misunderstanding of which sense of *animal* is being invoked—can lead to the kind of *reductio ad absurdum* that often undermines animal advocacy, at least when animal advocates are not preaching to the choir. It is relatively easy to explain why pigs and dogs deserve the same legal and moral consideration, although cultural biases make it much less easy to ensure that they actually receive equal treatment. The movie *Babe* to the contrary notwithstanding, even piglets have usually seemed less intrinsically appealing than puppies. Nevertheless resistance to acknowledging suine claims to humane treatment tends to rest on pragmatic (mostly economic) grounds. When, under the general "animal" rubric, claims to consideration are made on behalf of creatures less similar to people, resistance becomes stronger and more principled—the more so as the gap widens. If they are defended in the same terms as those used to defend our fellow mammals (or even our fellow vertebrates), the rights of lobsters, oysters, or termites offer ready targets for ridicule. (Of course, the distinction between the reasonable and the ridiculous is highly contingent, both culturally and historically. Two centuries ago Mary Wollstonecraft's *Vindication of the Rights of Women* was travestied on the grounds that if rights were granted to women, farmyard animals would be next in line.)

The most sweepingly inclusive (or powerfully reductive) categories thus make more sense for scientists than they do for scholars in the humanities and the social sciences. Biology has offered increasingly detailed and fascinating accounts of the genetic similarities that connect the smallest, simplest animals with the largest and most complex, and, indeed, that unify all the eukaryotes, whether animal, plant, or fungus. But such insights have little impact on everyday understanding and behavior at present, and their retroactive influence is still more limited. The study of human cultures, whether contemporary or historical, requires a focus that is at once larger and smaller. For understanding the relationships between people and other animals, the

Introduction

fact of similarity is important, but so also is the extent of similarity, which tends to be a matter of opinion or perception. It varies from place to place and from time to time. For example, although the general structure of mammalian taxonomy has remained reasonably constant for several centuries, Anglophones tend to feel closer to gorillas and chimpanzees now than they did in the late nineteenth century. The once common notion that dogs, or even horses, might bear a stronger resemblance to people in important ways has largely disappeared.

As the animal turn in scholarship breaks new ground, it also revisits perpetually unanswered questions. The standing of animals, even those closest to us, still presents vexed moral, legal, and political issues, and the range of possible positions is not very different from the range available in the nineteenth century. Nevertheless, consensus has shifted, at least to some extent, in the direction of consideration and respect. At the same time that this shift has offered a new way to interpret past history and culture, it has also presented significant new opportunities for misinterpretation. It is particularly difficult to avoid judging the past by the standards of the present. Such challenges are not unique to the scholarly study of animals, of course. They often emerge in the course of attempts to retrieve the history and cultural significance of marginalized human groups.

Paradoxically, however, the marginalization of animals, like that of humans, does not exclude the possibility of centrality. I began this introduction by saying that within my own experience as a scholar, which is charted by the essays collected in this volume, the study of animals has become more respectable and more popular within many disciplines of the humanities and the social sciences. It is, however, still far from the recognized core of any of them. It remains marginal in most disciplines, and (not the same thing) it is often on the borderline between disciplines. This awkward position or set of positions is also the source of much of its appeal and power. Their very marginality allows the study of animals to challenge settled assumptions and relationships—to re-raise the largest issues, both within the community of scholars and in the larger society to which they and their subjects belong.

Notes

1. Lévi-Strauss, *Totemism*, 89.
2. Carroll, *Annotated Alice*, 268–69.

3. Linnaeus, *Systema Naturae*.
4. Darwin, *Descent of Man*, 919.
5. For an extended discussion of eighteenth- and nineteenth-century hybrids and crossbreeds, see Ritvo, *Platypus and the Mermaid*, chap. 3.
6. White, *Account of the Regular Gradation in Man*, 34; Blumenbach, *Anthropological Treatises*, 73.
7. Tyson, *Orang-outang*, 2.
8. Quoted in Dudley Wilson, *Signs and Portents*, 56–57.
9. Isaac Watts, "Against Idleness and Mischief."
10. Kingsley, *Alton Locke*, 263–65.
11. On Koko, see the Gorilla Foundation website, http://www.koko.org/.

— I —

Sex and the Single Animal

In 1828, the author of an article in the *Farrier and Naturalist*—a livestock-oriented journal—asked, "What is . . . the part of the female in the great act of reproduction . . . ?" He answered his own question: "When the male predominates by his vigour, his constitution, and his health, she is limited, in some measure, to perform the same office that the earth does for vegetables . . . nothing more than a receptacle, in which are deposited the seeds of generation."[1] A few years later, William Youatt, the most distinguished British veterinarian of the early Victorian period and a prolific writer on domestic animals, recounted the following story to illustrate the need to control the imagination of "even so dull a beast as the cow": A "Cow chanced to come in season, while pasturing on a field . . . out of which an ox jumped, and went with the cow, until she was brought home to the bull. . . . The ox was white, with black spots, and horned. Mr. Mustard [the cow's owner] had not a horned beast in his possession, nor one with any white on it. Nevertheless, the produce of the following spring was a black and white calf with horns."[2] Early in the twentieth century, Judith Neville Lytton, a prominent if iconoclastic and combative member of the toy dog fancy, suggested the following remedy for barrenness in prize bitches: "In desperate cases . . . try the old recipe of breeding to a thorough cur. . . . If the bitch breeds to this connection . . . the next time the bitch is put to a thoroughbred dog she will almost certainly breed to him. . . . The more . . . highly bred the bitch is, the

---

"Sex and the Single Animal" originally appeared in *Grand Street* 7, no. 3 (Spring 1988): 124–39.

more likely this is to succeed."[3] Each of these statements depended on assumptions that are obviously false in the light of modern science and that could have been persuasively contradicted by the experience of contemporary experts. Yet it was the experts themselves who held these opinions (or believed these facts, as they would have put it), and who chose to express them in language conventionally used to describe human social intercourse.

Both the form and the content of these excerpts were determined by human gender stereotypes rather than by applied biology. This was the case even though animal husbandry was a quintessentially earthbound pursuit, constrained by physicality and detail on every side. It had no obvious connection to any arena of human social discourse. Certainly the techniques used to breed animals were very different from those used to breed people. The results of breeding were highly concrete, and they were usually presented to the general public in the stripped-down terms of cash value. Given all this, it would be reasonable to assume that breeders' understanding of their craft would demonstrate the determining power of empiricism rather than the pervasiveness of interpretation. Yet what they said about their cattle, sheep, and dogs was conditioned by their views about the nature of human beings, especially women.

By the nineteenth century, the breeding of pedigreed animals had become a highly specialized endeavor, whether it was carried on by professional agriculturalists interested in producing improved farm livestock or by self-professed amateurs who concentrated on dogs, cats, and smaller animals. Since the primary goal of all breeders was to produce superior young animals, their crucial focus was the selection of healthy and appropriately endowed parents for the new generation. Several factors encouraged them to be as pragmatic as possible in their matchmaking. In the first place, mistakes were easy to spot. Stringent standards existed for almost every kind of animal that was frequently bred, and these standards were widely disseminated in handbooks, prints, and periodicals, and vigorously enforced by show judges and by the marketplace in which animals were bought and sold by knowledgeable fanciers. (There were occasional exceptions to this rule of consensus and conformity, the most notable being the pig, which seemed so gross and amorphous that breeders had trouble figuring out what an ideal animal should be like.)[4] These frequently reiterated standards meant that the inferiority of the offspring of ill-considered pairings would sooner or later become obvious—perhaps at birth, but certainly by the time they reached maturity.

In addition, breeding was an expensive pursuit—the larger animals could cost hundreds and even thousands of pounds to purchase and then to maintain in an appropriate style. Any pregnancy risked the life of the mother, and each successful pregnancy consumed a significant portion of her reproductive potential. Owners of valuable female animals had to expend this limited resource very carefully.

The earliest and most celebrated achievement of English animal breeding was the modern thoroughbred racehorse, which appeared in the early eighteenth century as the result of an infusion of Arabian blood into native English equine stock. The merit of such horses was easily measured on the track. A secondary result of this development was the transformation of fox hunting, to which thoroughbreds that did not meet the highest racing standards were relegated. Their superior speed and jumping ability inspired the development of modern foxhounds, which were much more efficient than the motley animals they replaced.

By the middle of the eighteenth century, agriculturists were applying the techniques developed by racehorse breeders to farm livestock, with impressive increases in the sizes of cattle and sheep, most notably, but also of pigs and draft horses. The nineteenth century also saw an explosion in the diversity of fancy animals. Most modern dog breeds originated then, despite the claims of greater antiquity made by some aficionados; the same was true for cats, rodents, and the diverse strains of pigeons that Darwin studied. The record of serious animal husbandry in the eighteenth and nineteenth centuries thus seems to be one of straightforward, quantifiable, pragmatically oriented achievement. New methods were developed—albeit mostly through trial and error—and carefully applied, with predictably impressive results.

This was the way the master breeders themselves understood their accomplishments. Their published reflections on their craft suggested that the complex procedures they described could be rather mechanically applied either to the improvement of whole breeds by those at the forefront of husbandry or to the imitation of such results by breeders content to follow modestly in paths blazed by others. Thus one early Victorian manual for sheep breeders confidently associated the method with the result, asserting that "there cannot be a more certain sign of the rapid advances of a people in civilization and prosperity, than increasing attention to the improvement of livestock"; in a related vein, an earlier agricultural treatise had assured readers that "perfecting stock already well-bred is a pleasant, short and easy

Eighteenth-century racehorse, from Thomas Bewick, *General History of Quadrupeds*, 1824.

task."[5] At the end of the century, the author of a handbook for cat fanciers similarly suggested that good results would follow the methodical application of expertise: "[Mating] requires . . . careful consideration, and . . . experience and theory join hands, while the knowledge of the naturalist and fancier is of . . . superlative value."[6] Pedigree charts, the ubiquitous schematic representations of the results of animal breeding, corroborated this rather mechanical sense of what the enterprise involved.

When their discussions of animal breeding became more specific, however, the experts tended to backpedal. Neither method nor knowledge, even when they were operating on cats of impeccable pedigree and robust health, could assure "anything like certainty," according to the expert just cited.[7] Manuals for breeders of cattle, sheep, and horses often warned novices not to attempt to produce the kinds of animals that won prizes at national shows, because of the difficulty, risk, and expense involved. Thus, when closely scrutinized, animal breeding no longer seemed merely a mechanical procedure, but was implicitly redefined as a more ambiguous and impressionistic activity. And the more precisely the instructions were articulated, the more confusing they became. Often experts raised issues or offered advice that was irrelevant to the achievements of their stated aims, or even counterproductive. Old and widely recognized canards were ritually chewed over for decades after they had been persuasively discredited.

An explanation might stress the derivative nature of many of these works, which occupied the borderline between technical and popular writing, or it might focus on the conservatism inherent in many fields of applied technology. But the anomalous elements in the discourse of animal breeding can be most fully explained if that discourse is also viewed as an arena for the discussion of human gender issues. In a way it was an extremely obvious focus for such concerns—after all, the central task of breeders was to manage the sexuality of their animals. This task often posed challenges beyond merely deciding which ones to pair. The fact that the participants in this discourse were unaware of its double function merely allowed them to air their views and worries more explicitly.

In deciding which animals to pair, breeders selected parents on the basis of both their individual quality (that is, the extent to which they possessed the characteristics that were desired in their offspring) and a set of general notions about the way that such characteristics were transmitted. For most of the nineteenth century, there were few authoritative constraints on such ideas. Despite the claims of scienticity that agriculturists had made since the eighteenth-century vogue for improvement, few of them and even fewer breeders of small animals belonged to the scientific community. The works of popular natural history that they were most likely to encounter did not deal with the questions of reproductive physiology that engaged some elite biologists.

Even if breeders had been aware of the most advanced contemporary research on reproduction, they could not easily have applied its results to their enterprise. Although it was clear to scientists, as it was to breeders, that sexual intercourse was necessary if the higher animals were to reproduce, there was no expert consensus until late in the nineteenth century about why this was so.[8] That is, the modern understanding of the balanced contribution of sperm and egg to the development of a new organism was unavailable to eighteenth- and nineteenth-century animal breeders. Without this knowledge, they were free to predict and interpret the results of their breeding ventures with reference only to their own experience. That experience was vast and, indeed, considered extremely valuable by open-minded scientists like Charles Darwin.[9] It also turned out to include breeders' attitudes toward other people, as well as their observations of generations of animals.[10]

Many eighteenth- and nineteenth-century theories of reproduction presented it as the gradual enlargement and development of a tiny but com-

VICTORIA,
Roan, calved May, 30, 1840,

Bred by and the property of Mr. Forrest; got by Doctor (3096), d. (White Rose) by Bedford Junior (1701), g. d. (Moss Rose) by Romulus (1403), gr. g. d. (Vesta) by Isaac (1129), — (Rosabella) by Northern Light (1281), — (Rosabella) by White Comet (1582), — (Old Rose) by Cattley's Grey Bull (1798).

Produce in		Names, &c. of Produce.	By what Bull.	By whom bred.	Present Owners, &c.
1843, July 26, roan,	B. C.	Prince of Wales	Symmetry, 5389	Mr. Forrest	Mr. Forrest
1844, roan,	B. C.	(dead)		do.	

VICTORIA,
Calved in 1838,

Bred by Mr. J. Harrison, Lowfields; got by Gledhow (2062), d. (Linthorpe) by Ranger), g. d. by Meteor (431), gr. g. d. by Windsor (698), — by Suworrow (636).

Produce in		Names, &c. of Produce.	By what Bull.	By whom bred.	Present Owners, &c.
1843, July 2, light roan,	C. C.	Victoria	Donation, 5927	Mr. Jas. Thompson	Mr. J. Thompson

VICTORIA,
White, calved January 22, 1839,

Bred by Mr. Houldsworth, the property of Mr. Beauford; got by Coltness (3430), d. (Wharfdale Lady) by 2nd Hubback (1423), g. d. (Whiteface) by Frederick (1060), gr. g. d. (Western Lady) by Western Comet (689), — by Western Comet (689), — by Western Comet (689), — (Haughton) by a son of Favourite (252).

1842, May 27, roan,	C. C.	Princess Royal (dead)	3rd D. of Northumberland, 3647	Mr. Beauford	Mr. Beauford
1843, April 15, roan,	B. C.	(S.)	do.	do.	do.
1844, March 21, roan,	B. C.	(S.)	do.	do.	do.
1845, Feb. 15, roan,	B. C.	Prince of Wales	do.	do.	do.

VICTORIA,
Roan, calved in 1834,

Bred by Mr. Middleton, the property of Mr. Harvey, Tillygreig; got by a son of Emperor (1974), d. by a Bull of Mr. Alder's, of Hazelrig, g. d. bred by Mr. Weir, Goswick.

Produce in		Names, &c. of Produce.	By what Bull.	By whom bred.	Present Owners, &c.
1838, April 25, roan,	B. C.	Autocrat	Son of Archibald, 1652	Mr. Harvey	(slaughtered in 1842)
1839, June 14, white,	B. C.	The Parson	Mynheer, 7262	do.	Mr. Sellar
1840, May 17, red & white,	C. C.	Flirt	do.	do.	Mr. Copland
1841, April 19, roan,	C. C.	Sally	do.	do.	Mr. Sime
1843, March 20, white,	C. C.	Spotless	Van Amburgh	do.	Mr. Harvey
1844, March 5, roan,	B. C.	Sir Robert Peel	Premier, 6308	do.	Mr. Reith
1845, March 3, roan,	B. C.	The Wetherby Knight	Holkar 2nd, 7091	do.	Mr. S. Crompton

VICTORIA,
Red, calved in 1836,

The property of Mr. Moses; got by Plenipotentiary (2436), d. by Mameluke (2257), g. d. by Prime Minister (2454), gr. g. d. by Surprise (2716).

1839, roan,	C. C.	Virtue	Remus, 7417	Mr. Moses	Mr. Moses
1840, roan,	C. C.	Virgia	Panton Favourite, 4646	do.	do.
1841, roan,	B. C.	(S.)	Belvedere 4th, 3130	do.	do.
1843, roan,	B. C.	Virgil	William, 6723	do.	Mr. Livsey

* VICTORIA,
White, calved in May, 1839,

Bred by and the property of Mr. Wm. Raine; got by Pyramus (4853), d. (Young Denton) by Young Rockingham (2547), g. d. by Denton (198), gr. g. d. by Ladrone (353), — by Henry (301), — by Danby (190).

1843, June 1, white,	C. C.	White Rose	George	Mr. Raine	Mr. Wade, U.S.
1845, May 29, roan,	B. C.	King Charles	Pagan, 6268	do.	Mr. Raine

Coates's Herd Book, vol. 6, 1846.

plete seed, but scientists who adhered to this viewpoint were divided about whether that seed was contributed by the male or by the female. Animal breeders, however, were of one mind about this question. Many, like the author of the *Farrier and Naturalist* article quoted earlier, defined the female parent as a mere receptacle. One expert, faced with explaining why it was not "an easy thing to produce at once very perfect animals, provided that males of the right form could be obtained," preferred not to posit the mother as a significant source of variation. Instead he proffered as fact that "the offspring will, to a greater or lesser extent, partake of the form and structure of the grandparents [that is, the grandfathers]." And if such an absolute assertion of male dominance needed modification in view of the obvious tendency of young animals to resemble both their parents, breeding experts still clearly reserved the more vigorous genetic role for the stud. The imagery of activity and passivity remained useful in the modified case; it suggested, for example, that "the male gives the locomotive, and the female the vital organs."[11]

The preponderance of male influence never escaped the attention of writers on these subjects for long, even if they were momentarily diverted by the need to comment on females. For example, a reminder that "without first class females the descendants will not shine . . . in the showyard" was predictably accompanied by the acknowledgment that "it must not be forgotten that the male has most influence in breeding."[12] The only situations in which it was generally considered that the female might disproportionately determine the results of procreation were those that introduced a different, and also powerful, cultural construct. Social superiority—that is, in the terms of animal husbandry, a more distinguished pedigree—might tip the scales in the direction of the female. As Youatt pointed out, the influence of "a highly bred cow will preponderate over that of the half-bred bull."[13] Since such exceptional circumstances could result only from extreme negligence or ignorance on the part of the breeder, however, they did not have to be incorporated into received wisdom. In general, breeders were advised to proceed on the assumption that "not only . . . is the male parent . . . capable of most speedily improving the breed of livestock . . . , but . . . *the male is the parent,* from motives of sense and sound polity, which we can alone look to for the improvement of our breed."[14]

The principles that guided the production of livestock animals were routinely applied to pet species. The author of a late Victorian cat-breeding manual assured his readers that "the outward characteristics are in great measure transmitted by the male cat."[15] Nor did the advance of biological knowledge necessarily shake the faith of animal breeders in their time-tested principles. Instead, as biological knowledge became available, the jargon of science could be appropriated to the service of the conventional understanding of animal reproduction. Everett Millais, one of the most prominent dog fanciers of the late nineteenth century, translated it into the new terminology as follows: "that the male . . . does influence the epiblastic and mesoblastic structures largely, and all out of proportion to the female is undoubted."[16]

So powerful was the influence attributed to at least some males that it was believed to determine the character of offspring in whose conception they had had no part. That is, males might gain access to the reproductive organs through the eyes of receptive females, as well as in the ordinary way. As a result, breeders anxious to preserve the purity and the quality of their stock had to guard the minds as well as the bodies of their impressionable female animals from such undesirable approaches. It went without saying that females would be both unable and disinclined to resist them. That is, they could not be trusted with the preservation of their own virtue, even on the level of imagination. Mr. Mustard's cow, referred to earlier, offered an extreme example of the passive feminine susceptibility posited by this view of relations between the sexes. Even though the off-breed ox who jumped into her pasture when she was in heat was not even completely male, he apparently left his mark on the calf she subsequently conceived after intercourse with the bull selected by the owner.

If females of a relatively stolid species were so susceptible to the influence of random males, it was not surprising that female dogs, who were considered both more intelligent and more excitable, had to be guarded still more closely. Breeders agreed that the animals they termed maiden bitches were particularly vulnerable to such external stimuli, and they advised that "due influence should be exercised in the thorough isolation of bitches . . . or more than a temporary evil and disappointment may occur."[17] But more experienced bitches were also at risk, and beginning breeders were warned that "even very close intimacy between a bitch during oestrum and a dog she fancies may influence the progeny, although the dog has not warded her."[18] The struggles between bitches and breeders were described in terms that evoked the stubborn daughter in romantic narratives who refused to accept her father's choice of suitor. Hugh Dalziel, who wrote about a variety of Victorian dog breeds, once owned a Dandie Dinmont terrier whose wayward emotions made her useless for breeding; she "became enamoured with a deerhound, and positively would not submit to be served by a dog of her own breed."[19]

Such inclinations were fairly common among bitches, who were likely to implement them by stubbornly resisting their owners' prudent attempts to cloister them. The authors of manuals frequently warned novice breeders that bitches in heat would make unimaginably subtle and persistent attempts to escape from whatever quarters they were confined in. But it was necessary to persevere in thwarting them, because the stakes at risk in the preservation of female purity were high. A match with an inappropriate partner, especially in the case of a virgin animal, was held to have consequences far beyond the issue of that particular mating, as if a female's first sexual partner in some sense established a permanent proprietorship over her reproductive capacities. (See "Understanding Audiences and Misunderstanding Audiences," chapter 7 in this volume, for an extended discussion of telegony, or "the influence of the previous sire.")

In addition to jeopardizing the quality (or legitimacy) of future offspring, the tendency of female animals to follow their own sexual inclinations was perceived to pose less concrete but perhaps equally troublesome threats. Although an occasional authority might compassionately recommend indulging the desires of bitches that were "highly fed" and "living luxuriously, as a means of using up their excess stock of material," an interest in sex for its own sake rather than as a means to procreation was considered an indication of depraved character.[20] The behavior of bitches, in particular, confirmed the worst male fears about female proclivities. Although a single copulation might have sufficed for pregnancy, bitches would wantonly continue to accept new partners as long as they were in heat; connected with this unseemly pursuit of pleasure was a culpable indifference to its providers. One early-nineteenth-century sportsman complained that "no convincing proof of satiety is ever displayed . . . and she presents herself equally to all," with the result that the largest of her suitors "is generally brought into action." He noted with some satisfaction, however, that oversexed bitches might pay for their failure to prefer refinement to brute strength; many died while bringing forth too-large puppies.[21]

According to other authorities, however, this unmatronly behavior might harm the offspring rather than the mother. In the dog, as in some other animals that routinely give birth to multiple offspring, it is possible for a single litter to have more than one father. Breeding authorities referred to this phenomenon as *superfoetation,* a technical term that made it sound like an aberration or a disease. Even the possibility of this occurrence would jeopardize the pedigree of the resulting litter, and so aspiring breeders were strongly advised that "for at least a week after the bitch has visited the dog, the precautions for isolating her must not be relaxed, or all her owner's hopes may be marred."[22] But social stigma and unwelcome half-siblings were not the only ills that newly conceived puppies might sustain as a result of their mother's licentiousness. Dalziel suggested that during or after an unsanctioned second copulation "excessive pain, terror, or other strong emotions, may affect the unborn pups."[23]

In other species too interest in copulation for its own sake signaled the weakness of female character. A late-eighteenth-century agriculturist criticized the female ass for being "full as lascivious" as the male, which he claimed made her "a bad breeder, ejecting again the seminal fluid she has just received in coition, unless the sensation of pleasure be immediately removed by loading her with blows." He similarly condemned the sow, who "may be said to be in heat at all times; and even when she is pregnant she seeks the boar, which, among animals, may be deemed an excess." An edifying contrast, he pointed out, was offered by the demure behavior of cows, which, once pregnant, "will not suffer the bull to approach them."[24]

In males, however, eagerness to copulate was matter for praise. The he-goat was admired as "no despicable animal . . . so very vigorous . . . that one will be sufficient for above an hundred and fifty she-goats."[25] Breeders agreed that the only reason to curb the enthusiasm of studs was physical rather than moral, since too-frequent copulation was feared to undermine the constitution of both sire and offspring. Thus one authority on horses complained that "our best stallions . . . cover too many mares in one season; and this is the reason why they get so few good colts"; another advised against pasturing a stallion with a herd of mares because in this situation "in six weeks, [he] will do himself more damage than in several years by moderate exercise."[26] Similarly one expert on pedigreed dogs warned, "If you possess a champion dog . . . do not be tempted to stud him too much, or you may kill the goose which lays the eggs of gold. One bitch a fortnight is about as much as any dog can do, to have good stock and retain his constitution."[27]

Despite the need to practice such precise copulative accounting, the sexual management of male animals was much simpler than that of females. In the company of a suitable partner, bulls, stallions, dogs, boars, and rams ordinarily did what was expected of them. But even after breeders had presented their female in the required receptive and unsullied condition to a suitable male, their cares were not over. At that point the female might decide to exercise an inappropriate veto power, offering an unmistakable challenge to the authority of her owner. After all, in an enterprise dedicated to the production of offspring, too much reluctance was as bad as too little, and resistance to legitimate authority was as unfeminine as proscribed sexual enjoyment.

The terms in which breeders described such insubordination expressed not only the anger it provoked, but the extent to which that anger reflected worries about sexual subordination within their own species. Some categories of females were viewed with special suspicion. For example, bitches of the larger breeds, animals whose physical endowments commanded respect whether or not they were feeling refractory, had "to be taped or muzzled . . . to prevent either yourself or the dog from being bitten." Maiden bitches, too, were "generally a great annoyance from first to last." Their coyness might have to be counteracted with coercion, although breeders were cautioned to remember "not too much."[28] But almost any kind of bitch might evince reluctance when confronted with a prospective mate not of her own choosing, in which case she could be castigated as "troublesome," "morose," or even "savage."[29] The prescribed remedy was "hard exercise, until the bitch is thoroughly exhausted"; often this would "reduce a tiresome animal to submission." Bitches that refused to participate willingly at this point provoked their owners to severer measures—measures that again recalled the clichés of romantic fiction. They might be "drugged with some narcotic," or, in the most serious cases of insubordination, the male dog and the human breeder might cooperate in what was referred to as a "forced service."[30]

Such characterizations clearly reflected contemporary discourse about human gender, even though the breeders of domesticated animals were not explicitly aware of this subtext as they shared their practical wisdom and rehearsed their triumphs. Although it was sometimes echoed by the women who began to write about small-animal breeding in the late nineteenth century, this rhetoric expressed a distinctively masculine point of view. Occasionally, however, breeders encountered situations that required them explicitly to consider the relationship between their technical concerns and their ideas about human genders; that is, they had to accommodate human females as spectators of and, increasingly, participants in their activities. The rhetoric resulting from such encounters defined women rather differently than did the rhetoric embedded in breeding manuals and agricultural encyclopedias. Indeed, the rather straightforward if unacknowledged parallelism between female animals and female humans that distinguished the breeding literature conflicted with social conventions requiring that women, at least those considered respectable, be protected from any contact with taboo subjects. That is, in one case women were implicitly portrayed as creatures of wayward and demanding sexuality; in the other they were overtly described as creatures almost without a sexual nature.

For the point of connection between the two discourses was inevitably sex, a topic that was difficult for men to discuss with women and nearly impossible for women to discuss publicly at all. Although the public description of animal fancying, whether of livestock or of pets, always stressed the magnificence or beauty of individual specimens, it was an open secret that mating was the central preoccupation of fanciers. Animals were valued primarily for their procreative prowess. No matter how perfect they were according to the standards set by breed clubs (which did not, as a rule, mention intact reproductive organs, or, indeed, any such organs at all in their descriptions of points), neutered animals were either excluded from shows or shunted into separate classes. And even for intact animals, the proof of the pudding was in the breeding shed. No number of ribbons and trophies could redeem a champion who was unable to pass on the qualities for which he or she had been recognized.

Despite its incessant concern with an unspeakable subject, however, animal breeding was not a proscribed activity. On the contrary, prize livestock breeding had been a prestigious pursuit of the agricultural elite since the middle of the eighteenth century, and in the course of the nineteenth century the breeding of pet animals was very unexceptionally institutionalized. Both were objects of widespread and unembarrassed public curiosity. One reason for this may have been that the very structure of fancy breeding—characterized by meticulous record-keeping, carefully arranged matches, and constant human surveillance and even interference—allowed breeders to distinguish between the kind of sex with which they were involved and the kind that was potentially upsetting. The sexual behavior of their animals was, in very obvious and explicit ways, under control, which meant, among other things, cloistered or concealed.

Perhaps in order to defend the sanitized practices of their own animals,

"Hansom, Miss! Yes, Miss! Cattle or Dog Show?"

Punch cartoon, 1862.

breeders were as vociferous as any of their fellow citizens in criticizing the offenses of less well regulated creatures against public decency. Canine copulation was felt to be a particularly unedifying performance, and it provoked the most frequent criticism. (It was also true that the habits of dogs made them most liable to observation at such times; livestock were usually sequestered in the country, and cats conducted their business at night, out of human view if not out of human earshot.) The fact that mating dogs are physiologically unable to disengage meant that a flagrant display of sexuality was exacerbated by a flagrant display of insubordination. As one dog breeder objected, "where they are permitted to run about and appear in such a state before the habitations of the respectable . . . it is a most disgusting shameful spectacle . . . there is, perhaps, no nuisance that stands more in need of compulsive correction."[31]

The efficacy of this distinction between indecorous public mating and decorous private mating had its limitations, however. Once women actually joined the fancy, feminine modesty and ignorance could not be protected by a wall of bluster and denunciation. So saturated in sex was the breeding enterprise that when it first moved from the barn and pasture to the public arena in the form of the prize livestock exhibitions of the late eighteenth century, show organizers expressed surprise and concern at both the fact of women spectators and the numbers in which they appeared. By the early decades of the nineteenth century, the rosters of exhibitors occasionally included women, although women were never a significant presence among livestock fanciers. However, as the dog and cat fancies, which were less demanding financially and logistically, developed into popular middle-class pastimes, many women became breeders and exhibitors. Although they tended not to take conspicuous or authoritative roles in the most conservative fancying institutions—for example, as officers of the major clubs or as judges at major shows—they were active participants in the life of their avocation. They were sufficiently numerous that in the 1890s a periodical directed only to them—the *Ladies' Kennel Journal*—flourished briefly. And while these women were occasionally castigated as unfeminine by their male colleagues, the epithet was ordinarily intended to call their attractiveness rather than their modesty into question.

The solution to this problem was worked out on the level of rhetoric, since the level of action offered little scope for compromise. In order to accommodate the delicate sensibilities of both women and the men who had to talk to them about such uncomfortable topics, an elaborate vocabulary of euphemisms was developed. Thus the *Fox Terrier Chronicle,* the first periodical devoted to a single dog breed, reported on matings under the rubric "visits." In this terminology, cat owners were advised that females usually "insisted on matrimony" before they were twelve months old; the union of a retriever with a pointer was termed a "mésalliance"; and the owner of a stud bulldog announced that he was available to "serve a few approved bitches."[32] The straightforward Judith Neville Lytton excoriated the prissiness of some of her fellow fanciers, who insisted on referring to lady dogs and gentleman dogs, but she too resorted to euphemism when the going got rough. For example, she referred circumspectly to the deceitfulness of male dogs, "which have an active dislike to any female which is in a condition to breed."[33]

This rhetoric of accommodation and compromise did not have much effect on the traditional discourse of animal breeding, which after all reflected much more accurately what really went on. Breeders who wrote about livestock animals for an overwhelmingly masculine audience made no concessions at all. And the dog fanciers who purged their vocabularies in deference to feminine sensibilities may well have felt put upon. Certainly many of their colleagues, especially sportsmen with a predilection for hunting or fighting dogs, complained bitterly about the feminization of their pastime. (There were no such complaints about the cat fancy, because from the beginning it had been as much a female as a male preserve. This probably reflected the fact that cats were more difficult to manipulate than dogs in every way, including both behavior and genetics.) And in fact, whether or not it provoked resistance, there would have been no reason to expect this feminized discourse to alter the traditional ideas of experts on breeding. For along with much information about the habits and inclinations of domesticated animals, the literature of breeding also presented a fundamentally male view of the relations between the sexes, a view determined by the notion that women were in greater need of control than of protection.

Notes

1. M. Godine, "Comparative Influence of the Male and Female in Breeding," *Farrier and Naturalist* 1 (1828): 468.

2. Youatt, *Cattle*, 523. It should be remembered that oxen are castrated animals.

3. Lytton, *Toy Dogs and Their Ancestors*, 194–95.

4. The struggles of breeders to determine the ideal form of the pig is elaborately chronicled in Wiseman, *History of the British Pig*.

5. Blacklock, *Treatise on Sheep*, 67; John Lawrence, *General Treatise* (1805), 30–31.

6. Harrison Weir, *Our Cats and All About Them*, 96.

7. Ibid.

8. Farley, *Gametes and Spores*, chaps. 1 and 2. See also Churchill, "Sex and the Single Organism."

9. James A. Secord has discussed the contribution of pigeon fancying to Darwin's theory of evolution in "Nature's Fancy." Darwin continued to make use of information supplied by animal breeders after he wrote *On the Origin of Species,* most notably, in *The Variation of Animals and Plants under Domestication.*

10. Exposure to scientific data and theory would not necessarily have made any difference. At least some Victorian scientists shared breeders' inclination to identify women with other mammalian females, eagerly but without much evidence conflating the human menstrual cycle with the oestrus cycle of dogs and cattle. Laqueur, "Orgasm, Generation," 24–25.

11. "The Physiology of Breeding," *Agricultural Magazine, Plough, and Farmers' Journal,* June 1855, 17.

12. M'Combie, *Cattle and Cattle-Breeders*, 153.

13. Youatt, *Cattle*, 524.

14. John Boswell, "Essay upon the Breeding of Live Stock, and on the Comparative

Influence of the Male and Female Parents in Impressing the Offspring," *Farmer's Magazine*, n.s., 1 (1838): 248.

    15. Jennings, *Domestic and Fancy Cats*, 45.
    16. Millais, "Influence," 153.
    17. Vero Shaw, *Illustrated Book of the Dog*, 525.
    18. Dalziel, *Collie*, 48.
    19. Dalziel, *British Dogs*, 462–63.
    20. Dalziel, *Collie*, 41.
    21. Taplin, *Sportsman's Cabinet*, 27–28.
    22. Vero Shaw, *Illustrated Book of the Dog*, 524.
    23. Dalziel, *Collie*, 48.
    24. Mills, *Treatise on Cattle*, 271, 310, 401.
    25. Ibid., 387.
    26. Hanger, *Colonel George Hanger, to All Sportsmen*, 47.
    27. Stables, *Practical Kennel Guide*, 125.
    28. Ibid., 123–24.
    29. Vero Shaw, *Illustrated Book of the Dog*, 523.
    30. Davies, *Kennel Handbook*, 66.
    31. Taplin, *Sportsman's Cabinet*, 27–28.
    32. Simpson, *Book of the Cat*, 341; "Kennel Advice," *Dogs* 1 (1894): 2; *Sportsman's Journal and Fancier's Guide* 9 (1879): 2.
    33. Lytton, *Toy Dogs and Their Ancestors*, 177, 195.

— 2 —

Learning from Animals

Natural History for Children in the Eighteenth and Nineteenth Centuries

The first zoological book intended for English children, *A Description of Three Hundred Animals,* appeared in 1730. Published by Thomas Boreman, it was part of a mid-eighteenth-century boom in juvenile literature, created by publishers rushing to cater to a market that had been virtually nonexistent before 1700. Because both the authors and the purchasers of children's books understood them primarily as educational tools, not as instruments of entertainment, it is not surprising that the natural world, especially animate nature, was quickly recognized as a source of useful information and instructive moral precepts.[1] By 1800, according to one bibliographer's count, at least fifty children's books about animals, vegetables, and minerals had been published.[2]

In the middle of the eighteenth century, knowledge about nature was accumulating rapidly. Natural history had become both a prestigious scientific discipline and a popular avocation.[3] An eager adult public awaited the dissemination of information collected by Enlightenment naturalists. Some had the training, patience, and money to appreciate such focused and elaborate treatments as William Borlase's *Natural History of Cornwall* (1758) or Thomas Pennant's *Arctic Zoology* (1784–87). But most awaited the popular distillations of such works. The versatile Oliver Goldsmith provided one of the most successful, an eight-volume compilation entitled *An History of the Earth and*

"Learning From Animals: Natural History for Children in the Eighteenth and Nineteenth Centuries" originally appeared in *Children's Literature* 13, no. 3 (1985): 124–39 (Copyright © 1985 The Children's Literature Foundation, Inc.), and is reprinted with permission of the Johns Hopkins University Press.

Animated Nature (1774). He was plundered in his turn by several generations of literary naturalists eager to supply the popular demand, including many authors who targeted the growing juvenile audience.

Although natural history was a new literary genre in the eighteenth century, animals were hardly new literary subjects. They figured prominently in Aesop's fables, which were frequently used as school texts in the sixteenth and seventeenth centuries. The fables, however, were not really about animals. As Thomas Bewick, a distinguished illustrator and publisher of animal books, explained in the preface to his 1818 edition of *The Fables of Aesop*, they "delineate the characters and passions of men under the semblance of Lions, Tigers, Wolves and Foxes."[4] Nevertheless, because the animals were supposed to bear some temperamental resemblance to the human characters they represented, the fables have always been perceived as animal stories as well as moral tales.

But fables exerted only an oblique influence on natural history writing. The impact of the bestiary tradition, which also had classical roots, was more direct and definitive. Bestiaries were illustrated catalogues or compendia of actual and fabulous animals. They can be regarded as forerunners of natural histories, sharing the same purpose—to describe the animal world—but adumbrating a different point of view. In Latin versions they were widely disseminated across Europe in the Middle Ages.[5]

The fruits of this tradition had been distilled for English readers early in the seventeenth century. Edward Topsell's massive, densely printed *Historie of Foure-Footed Beastes* (based on Konrad von Gesner's five-volume *Historia Animalium*, which had been published half a century earlier) described each animal emblematically, detailing its "virtues (both naturall and medicinall)" and its "love and hate to Mankind."[6] The information, which was miscellaneously gathered from ancient authorities, modern travelers' tales, and unattributed hearsay, could better be characterized as lore than as scientific data. Nevertheless, Topsell's collection exerted a strong influence on at least the form of natural history books well into the eighteenth century.

Like its manuscript predecessors, Topsell's *Historie* was intended for adults, but its bizarre stories and illustrations must also have been attractive to children. Perhaps for this reason the authors of the first natural history books for children mined it especially heavily. In so doing, however, they transformed the traditional genre of the bestiary in ways that reflected the concerns of their own age.

"The Dog and the Shadow," from Thomas Bewick, *Select Fables, with Cuts,* 1820.

Thomas Boreman's *Description of Three Hundred Animals* had been recognized as the first animal book aimed at children, because the preface announced that it was intended to "introduce Children into a Habit of Reading."[7] Without this clue, it might have been difficult to tell. In many cases, the material presented in animal books written for children in the late eighteenth and early nineteenth centuries did not distinguish them from works designed for an adult audience. Small size often indicated a book intended for small readers. For example, *The Natural History of Four-footed Beasts,* published by Newbery in 1769, measured approximately 2¾″ × 4⅛″ and had a tiny illustration (rather crude and unrealistic, with the animals sporting eerily human expressions) for each entry. T. Teltruth was the pseudonymous author, and the book was clearly meant for children. Yet the text showed no sign of special adaptation. The print was small, and the multipage entries included such oddly selected tidbits as that the flesh of the tiger "is white, tender, and well tasted" and that jackals "howl in the most disagreeable manner, not unlike the cries of many children of different ages mixed together."[8]

Some authors did adapt their material to a juvenile audience. For example, *A Pretty Book of Pictures for Little Masters and Misses, or Tommy Trip's History of Beasts and Birds,* which was first published about 1748 and reprinted into the nineteenth century, offered some one-page descriptions of the animals,

Illustration from Thomas Boreman,
Description of Three Hundred Animals, 1736.

each introduced by a doggerel quatrain. The anonymous author culled the standard authorities carefully for information that children would find interesting, appealing, and comprehensible. Thus the baboon was evoked in vivid physical detail—rough skin, black hair, large teeth, and bright eyes—and its proclivities for fishing and mimicry illustrated within a brief paragraph.[9] Most authors, however, were more concerned with the baboon's moral than with its physical character. Following Boreman, they spent several pages castigating baboons as ugly, surly, and disgusting, describing how troops of baboons attacked people. Throughout the eighteenth century, purchasers of children's books could choose between relatively materialistic and relatively moralistic approaches to the animal kingdom. *A Pretty Book of Pictures* and *The Natural History of Four-footed Beasts* coexisted for decades on the list of Newbery, the leading publisher of children's books.

Even as they catered to a distinctively eighteenth-century thirst for knowledge, these first children's natural history books recalled their medieval roots. Although he claimed that his information was "extracted from the best authors," Boreman crammed *A Description of Three Hundred Animals* with legendary material. Along with the lion, the bear, the ox, and the beaver appeared a host of mythical beasts. The entry on the unicorn acknowledged that it was "doubted of by many Writers," but no skepticism was expressed about the Lamia, with "Face and Breasts like a very beautiful Woman . . . hinder Parts like a Goat's, its forelegs like a Bear's; its Body . . . scaled all over," or the similarly patchwork "Manticora," "Bear-Ape," and "Fox-Ape."[10] Of the "Weesil," an animal native to Britain and familiar to most country people, Boreman reported that it was "said to ingender at the Ear, and bring forth [its] Young at the Mouth."[11]

Although Boreman's successors tended to borrow their information from more reliable sources, they nevertheless perpetuated the bestiary format. Animals were catalogued one by one, and each entry was introduced by an illustration, which was at least as important as the text in attracting an audience. In most cases, as in the bestiaries, the entries seemed randomly ordered, after an initial appearance by the king of beasts. Thomas Bewick's *General History of Quadrupeds* (1790) was unusual in using, as had Goldsmith, a rough semblance of Linnaean categories—such as the horse kind, the hog kind, and the "sanguinary and unrelenting" cat kind.[12] More typical was *The British Museum; or Elegant Repository of Natural History*, by William Holloway and John Branch, which put wolves next to elephants and peccaries next to opossums. The "guide to the zoo" was a nineteenth-century variation on this theme that presented the animals according to either the layout of the zoo in question or the attractiveness of the different exhibits to visitors.[13]

Within this traditional format, however, the kind of information presented had changed significantly. Even Boreman's rather fantastic work appealed to the newly scientific temper of his age. The bestiaries had described animals as figures in human myths or allegories of human concerns. Boreman assumed that his readers were interested in quadrupeds for their own sake, just because they existed as a part of external nature. He asked not "What do they mean?" but "What are they like?" His entries, like those of most of his successors, focused on the animal's mode of life, physical appearance and abilities, temperament, moral character, and possible utility to man.

Because natural history was perceived to be intrinsically interesting to children, books about it were ideal didactic instruments. The educational theories of John Locke, at once more pragmatic and more human than their predecessors, had redefined the function of early education. The purpose of children's books was to entice them to learn rather than to force them.[14] Thus Boreman suggested that his subject matter was preferable to that ordinarily offered by introductory readers, which was "such as tended rather to cloy than Entertain."[15] Or, as the advertisement for *The Natural History of Beasts* (1793), attributed to Stephen Jones, proclaimed, "The study of Natural History is equally useful and agreeable: entertaining while it instructs, it blends the most pleasing ideas with the most valuable discoveries."[16] This was especially important for middle-class children, who were the main audience for juvenile books, and whose parents, it is safe to assume, were eager for them to succeed in an aggressive commercial society.[17] By seducing children into frequent and careful reading, natural history books helped instill future habits of energetic and studious application.

If the study of nature in general was instructive, the study of the animal creation was more rewarding still. Quadrupeds or beasts, in particular, frequently received special attention. (Both terms were used in the eighteenth and nineteenth centuries as synonyms for *mammals,* which was considered alarmingly pedantic by adults as well as children.)[18] Their greater similarity to man rendered them both more interesting than and intrinsically superior to other animals.[19] In addition, they were easier to observe and to interact with; unlike birds, fish, and reptiles, they occupied more or less the same space as man and, as one pragmatic author pointed out, "cannot easily avoid us."[20]

A scientific understanding of the animal kingdom was thought to enhance not only studious habits but also religious feeling. According to Holloway and Branch, "No other [human pursuit] excites such proper sentiments of the being and attributes of God."[21] Two decades later, the anonymous author of *The Natural History of Domestic Animals* was more explicit about how these effects were produced: "Whilst we observe, therefore, so many instances of the Almighty's wisdom and goodness, in these which are his creatures, let us humbly and gratefully acknowledge him as the source of all our happiness."[22] This connection persisted even after Darwin had put the scientific order of creation at odds with the religious one. As late as 1882, Arabella Buckley

The dog-faced baboon and the purple-faced monkey,
from Thomas Pennant, *History of Quadrupeds*, 1793.

claimed that the purpose of her strongly evolutionary introduction to vertebrate biology, *The Winners in Life's Race,* was to "awaken in young minds a sense of the wonderful interweaving of life upon the earth, and a desire to trace out the ever-continuous action of the great Creator in the development of living beings."[23]

Understanding the order of creation would also encourage children to treat animals with kindness. Late-eighteenth-century moralists were almost obsessively concerned with children's propensity to torture insects, birds, and small domestic animals, as much because it was a prognostication of adult behavior toward fellow humans as on account of the animal suffering it caused. The main crusaders against this kind of cruelty were sentimental fabulists like Sarah Kirby Trimmer and Samuel Pratt.[24] Natural history writers shared the concern of the fabulists, but they addressed their readers' heads as well as their hearts. Thus in *The Rational Dame,* Eleanor Frere Fenn used the results of scientific observation to demonstrate that although inferior in rank to man, animals shared his ability to feel—that "man is the *lord,* but ought not to be the *tyrant* of the world."[25]

Thus the study of natural history was morally improving. But it did not separate children from more practical considerations. If benevolence and piety were intrinsically laudable, they were also associated with more tangible rewards. God's order itself was understood to be good because it benefited man. Fenn found in the animal world "the most evident appearances of the Divine Wisdom, Power, and Goodness," one example of which was "how wisely and mercifully it is ordained, that those creatures that afford us wholesome nourishment, are disposed to live with us, that we may live on them."[26] The author of *The Animal Museum* appealed first to the highest moral authority in urging children to treat animals "as the property of our common Creator and Benefactor, with all the kindness their nature is capable of receiving." Then he suggested an additional motive: "This conduct is not only our duty, but our incentive; for all the animals domesticated by man or that come within the sphere of his operations are sensible of kindness, and but few are incapable of some return."[27]

In addition to direct moral lessons, children's books about animals were crammed with information that might also have desirable moral consequences. Thomas Varty's *Graphic Illustrations of Animals* consisted of a series of enormous colored cartoons, each devoted to a single animal or group of animals. The one that displayed "The Bear and Fur Animals," for example, featured a central illustration of bears, beaver, lynx, and mink in a northern pine forest, flanked by smaller pictures of the animals transformed into such useful objects as winter coats, soldiers' hats, royal regalia, perfume, paintbrushes, and food (bear tongues and hams were considered delicacies). Understanding how useful animals were—that they constituted "the life of trade and commerce, and the source of national wealth . . . the cement of human society"—would impress the mind of a child reader with the improving emotions of "gratitude, admiration, and love."[28]

While being uplifted, however, he would also be edified. According to Varty, the "graphic illustrations" would allow the child to form a just estimate of the "intrinsic value of each creature," independent of sentimental considerations such as beauty or amiability. The practical value of compilations that surveyed only the domestic animals, "which human perseverance has reclaimed from wildness and made subservient to the most useful purposes" (often the same compilations that stressed the importance of human treatment most heavily), was considered so obvious as to require no further explanation. And even information about more exotic animals might come in handy. Thus the camel could substitute for the horse in a desert, as the goat could replace the sheep in harsh climates.[29]

Wild animals too might serve practical human purposes. They could be killed for their skins and horns, or for their flesh; they could be tamed as pets or as performers. A shrewd eye could recognize which wild animals were likely candidates for domestication. The zebra, for example, recommended itself as a carriage animal by its beauty and its similarity to the horse; in the view of one naturalist, "it seems formed to gratify the pride of man, and render him service." That it had not yet been tamed by the human inhabitants of its native savanna was ascribed to their lack of information and enterprise: they had "no other idea of the value of animals of the horse kind, but as they are good for food."[30] Well-instructed adventurers would neglect no such opportunity. Thus, learning about animals could help children be good, and it could help them do well.

The most important lesson taught by animal books was less directly acknowledged by their authors. This was a lesson about the proper structure of human society. Quadrupeds occupied a special position in relation to man, a position symbolized but not completely described by their biological closeness. (This closeness, which was recognized long before Darwin, did not imply any evolutionary connection.) Both religion and experience taught that they had been created for human use; some kinds even seemed to seek, or at least to accept without protest, human companionship and exploitation. The attraction was reciprocal; as Mary Trimmer put it, quadrupeds were unlike "birds, fishes, serpents, reptiles, and insects" in the greater extent to which "their sagacity and constancy of affection excite our observation and regard." People and quadrupeds seemed to understand each other. In all, "their circumstances bear some analogy to our own."[31]

By learning about animals children could also learn about mankind. The animal kingdom, with man in his divinely ordained position at its apex, offered a compelling metaphor for the hierarchical human social order, in which the animals represented subordinate human groups. Embodying the lower classes as sheep and cattle validated the authority and responsibility exercised by their social superiors. Embodying the lower classes or alien groups as dangerous wild animals emphasized the need for their masters to exercise strict discipline and to defend against their depredations. These identifications were nowhere explicitly stated, but they constantly informed the language used to describe the various animals. In addition, they were implicit in the system of values that determined the moral judgment pronounced upon each beast.

What was explicitly stated was the inferiority of animals to man. For this reason the metaphorical hierarchy remained incomplete; animals never exemplified the best human types. But the sense of human dignity that barred animals from realizing, even figuratively, the highest human possibilities made them particularly appropriate representatives of the less admired ranks and propensities. If animals carried the message—if it were not completely clear where natural history ended and social history began—it might be easier to teach children unpalatable truths about the society they live in.

The dividing line was reason, "the privilege of man." Although the behavior of some animals "often approaches to reason," according to the author of *Animal Sagacity,* it never crossed the impenetrable boundary; "men weigh consequences . . . animals perform their instinctive habits without foreseeing the result."[32] This distinction justified man's domination of animals, both pragmatically and in principle. According to Mary Trimmer, "While man is excelled in strength, courage, and almost every physical excellence, by some one or other of the animal creation, he is yet able, aided by intellect, to subject to his own uses the very powers, which, properly directed, might greatly injure, if not destroy him." And in this case, at least, might made right. The "subserviency" of quadrupeds "to our comforts and wants" was therefore "manifest."[33]

Even the sentimental fabulists were firm about the line separating man and beasts, a line that placed certain ineluctable limits on the obligation to be kind to them. It was, for example, permissible to exploit them economically in all the usual ways.[34] In no case, according to these earnest didacticists, should concern for animals eclipse concern for other human beings. Although during most of Sarah Kirby Trimmer's *Fabulous Histories,* the virtuous Mrs. Benson concentrated on teaching her children to be kind to rather humanized animals, she also included a salutary lesson on the dangers of excessive fondness. The foolish Mrs. Addis neglected her children while doting on her birds, squirrel, monkey, dog, and cat. Eventually the animals died, and her children turned out badly, leaving her to an old age of loneliness and regret.

The need to distinguish appeared most clearly when the resemblance was most striking. Descriptions of apes and monkeys often vacillated between admiring recitals of their resemblances to man and firm denials of their closeness. Orangutans were said to walk erect, to build huts, to attack elephants with clubs, and to cover the bodies of their dead with leaves and branches. One status-conscious ape, bound for England by ship, expressed his sense of kinship with mankind by embracing the human passengers whenever possible and snubbing some monkeys who were also aboard. But, as the author of *The*

Animal Museum noted, these similarities were "productive of . . . few advantages": orangutans could not talk or think.[35]

Monkeys illustrated the dissociation of physical and intellectual qualities still more satisfactorily. Despite occasional reports of their extraordinary sagacity—one Father Carli, a missionary, found the monkeys more tractable than the human residents of Angola—they were usually characterized as "mischievous" at best, "filthy" and "obscene" at worst.[36] Yet as T. Teltruth detailed in *The Natural History of Four-footed Beasts,* they resembled humans closely in the face, nostrils, ears, teeth, eyelashes, nipples, arms, hands, fingers, and fingernails. This similarity, however, turned out to be completely superficial. Teltruth reassured his readers that monkeys "if compared to some quadrupeds of the lower orders, will be found less cunning, and endowed with a smaller share of useful instinct."[37]

In the case of quadrupeds, zoology was destiny. Their inferior mental capacities dictated their subordination to man. As with people, subordination was routinely expressed in terms of servitude; natural history writers urged children to wonder what use the various beasts could be to them. Although some wild animals could be harvested, the most useful species were those that "man has subjected to his will and service."[38] So domestic animals, described in terms that suggested human domestics, provided the model by which other animals were judged: "They seem to have few other desires but such as man is willing to allow them. Humble, patient, resigned, and attentive, they fill up the duties of their station, ready for labour, and satisfied with subsistence."[39] By a somewhat circular calculation, animal intelligence or sagacity was equated with virtue. Like the best human servants, the best animals understood their obligations and undertook them willingly; the worse were those that not only declined to serve but dared to challenge human supremacy.

For this reason, the most appreciated domestic animals were not the sheep, "the most useful of the smaller quadrupeds," or even the ox (the term used generically for cattle), whose "services to mankind are greater than those of sheep, for . . . they are employed . . . as beasts of draught and burden."[40] Occasionally these beasts might show some understanding of their special bond with mankind—for example, a ewe who led a girl to a stream where her lamb was drowning or a bull who showed gratitude to a man who had protected him from lightning.[41] And it was pleasant (especially in contrast to "the savage monsters of the desert") "to contemplate an animal designed by providence for the peculiar benefit and advantage of mankind."[42] Nevertheless, cows, on the whole, were merely "gentle," "harmless," and "easily governed by Men," and sheep, though "affectionate," were "stupid";[43] both kinds were the equivalent of mindless drudges.

The services of animals able to understand their subordinate position and accept its implications were valued more highly. The horse was repeatedly acclaimed as "noble." In part this accolade reflected its physical magnificence, for it was "more perfect and beautiful in its figure than any other animal" and "adapted by its form and size for strength and swiftness."[44] Even more worthy of admiration, however, was the fact that, although "in his carriage, he seems desirous of raising himself above the humble station assigned him in the creation," the horse willingly accepted human authority.[45] "With kind treatment," according to one appreciative writer, it would "work till it is ready to die with fatigue."[46] Horses were affectionate creatures—there were many stories of their attachment to stablemates and farmyard animals of different species, as well as to people—and their understanding was, at least in the opinion of some admirers, "superior to that of any other animal."[47] This perspicacity produced "a fear of the human race, together with a certain consciousness of the services we can render them."[48]

Even more eager and aware in accepting the bonds of servitude was the dog, the favorite species of almost all the writers who described the animal kingdom for children. Like the horse, its only competitor for the highest appreciation, the dog was said to combine extreme sagacity (the term regularly employed by those reluctant to assign "intelligence" to animals) with affection and obedience. According to *The Natural History of Beasts,* the dog was characterized by "affectionate humility. . . . His only aim is to be serviceable; his only terror to displease."[49] Stories of dogs who had preserved their masters' lives and property were so routine that it was worthwhile recounting only those in which the animal had displayed unusual devotion or shrewdness, such as when a ship's dog saved the whole crew by warning them that the hold was filling with water or when an alert watchdog caught a human fellow servant stealing corn.[50] Such demonstrations made the dog "the most intelligent of all known quadrupeds"; in addition it was "the most capable of education."[51] It was "the only animal who always knows his master, and the friends of the family."[52] The dog's mental powers were such that "in the rude and uncultivated parts of the earth, he might, in point of intellect . . . be placed almost upon a footing with his master," yet it never showed dissatisfaction with its subordinate rank. It wanted nothing more than to be "the friend and humble companion of man."[53]

Some domestic animals had trouble meeting even the minimal standards of obedience set by sheep and cattle, let alone the high standards of cooperation set by the dog and the horse. Like disrespectful underlings, they did not adequately acknowledge the dominion of their superiors. The pig, for example, despite its incontestable value as a food animal—"ample recompense . . . for the care and expense bestowed on him"—was routinely castigated as stupid, filthy, and sordid, seeming "to delight in what is most offensive to other animals."[54] Pigs were defective in morality as well as in taste. Sows were accused of devouring their own young, which in turn scarcely recognized their mother.[55] Naturally, they did not recognize their human caretakers. Even physically, they were less responsive to the guiding hand of man; according to one writer, "The hog seems to be more imperfectly formed than the other animals we have rendered domestic around us."[56]

Although the cat could not have been more different from the pig in its beauty and cleanliness, it had similarly resisted human efforts to mold it physically. Nor did it seem disposed to accept other forms of domination. It served man by hunting and thus did not depend on people for sustenance. It was suspected of having "only the appearance of attachment to its master," really either "dreading" him or "distrusting his kindness"; people feared that "their affection is more to the house, than to the persons who inhabit it."[57] It was considered faithless, deceitful, destructive, and cruel; it had "much less sense" than the dog, with which it was inevitably compared; and, in all, it was only "half tamed."[58] Its diminutive resemblance to the lioness and the tiger provoked many uneasy remarks.

If domestic animals symbolized appropriate and inappropriate relations between human masters and servants, the lessons to be drawn from wild animals were more limited. This may explain the surprising extent to which zoological popularizers neglected exotic wild animals in favor of familiar domestic beasts. For example, in Bewick's *General History of Quadrupeds*, which appealed to both children and adults, the briefest entries were less than a page long and the standard entry for a significant wild animal (one that was reasonably well known and about which some information was available) was from five to nine pages. Yet thirteen pages were devoted to the horse, fourteen to the ox, seventeen to the sheep, eleven to the goat, eleven to the hog, and thirty-nine to the dog. The only wild animals to receive comparable attention were the elephant and the squirrel, which could be measured by the standards set by domestic animals. The elephant had been semidomesticated in India. Although it did not breed in captivity, it was easily tamed, in which condition it was docile, mild, and an "important auxilliary to man."[59] As a result, it was also characterized as noble, friendly, courteous, and sagacious.[60] Like elephants, squirrels were easily tamed. Unlike elephants, they were frequently kept as pets by English children, who might learn from them to be "neat, lively . . . and provident."[61] In any case, their willingness to abandon their "wild nature" for domesticity had made them "fine" animals, "universally admired."[62]

The descriptions of many other wild animals were neutral in tone. Writers were unable to muster much enthusiasm about the fact that exotic animals like the raccoon and the capybara (an enormous rodent) were tameable or that the endless variety of deer and antelope encountered in every newly explored territory could all be eaten.[63] Some speculations showed a limited sympathy for strange creatures; for example, the author of *The Natural History of Animals* remarked of the sloth that "though one of the most unsightly of animals, it is, perhaps, far from being miserable."[64] For the most part, however, animals were not even important enough to merit a moral judgment unless they somehow influenced human experience. Thus the rhinoceros, the giraffe, the hippopotamus, the badger, and the camel were dismissed as simply inoffensive.[65]

"The Improved Cart-Horse," from Thomas Bewick, *General History of Quadrupeds*, 1824.

Beasts of prey were seldom dismissed in this way. Their carnivorous way of life disposed them to challenge man rather than to serve or flee him; they were rebels who refused to accept his divinely ordained dominance. Natural history books for children therefore tended to present them as both dangerous and depraved, like socially excluded or alien human groups who would not acknowledge the authority of their superiors. (Sometimes this analogy was made explicit, as when the author of *The Natural History of Beasts* noted that "in all countries where men are most barbarous [that is, Africa], the animals are most cruel and fierce."[66]) Even small creatures that could not directly defy human power were castigated for their predatory propensities. The weasel, for example, was "cruel, cowardly, and voracious."[67] If such animals could not be controlled, they might have to be exterminated. "However much we detest all cruelty to the brute creation," intoned the author of *The Animal Museum,* the fox "is so destructive to the property of the farmer . . . that his destruction is absolutely necessary."[68]

Large, powerful animals were, naturally enough, even more threatening, and, with one exception, they were described as unmitigatedly wicked. The exception was the lion, whose prestige as the king of beasts (lingering from the medieval bestiaries) was enhanced by its contemporary function as the emblem of British power. Although it was acknowledged to be dangerous and powerful, it was praised for its generosity and magnanimity in using its strength.[69] It attacked bravely, from the front, and never killed unless it was hungry. Most important, the lion respected man. It had learned to fear human power, and according to the African explorer Mungo Park, whose travels were available in a special children's edition, it would "not offer violence to a human being, unless in a state of absolute starvation."[70] (At least not to Europeans—another naturalist, perhaps more learned but with less hands-on experience, opined that "the Lion prefers the flesh of a Hottentot to any other food."[71])

The tiger was the reverse of the lion in every way, the epitome of what man had to fear from the animal kingdom. If the lion was the judicious king of beasts, the tiger was the evil, usurping despot. Its beauty cloaked "a ferocious and truly malignant disposition."[72] Indeed, the tiger's appearance so misrepresented its character that Holloway and Branch warned their young audience that "providence bestows beauty upon so despicable an animal to prove, that when it is not attached to merit, it neither deserves to be estimated or prized."[73] It was cruel and greedy, interrupting a meal of one carcass to kill another animal or slaughtering an entire flock and leaving them dead in the field.[74] Like the wolf and some other big cats, the tiger was often called "cowardly," which apparently meant unwilling to face men with guns.[75] Nevertheless, it did not fear man and refused to respect him. The authors of *The British Museum* used the language of redemption to lament that "no discipline can correct the savage nature of the tiger, not any degree of kind treatment reclaim him."[76]

The ultimate index of the tiger's unregeneracy was its fondness for human flesh. Not only was it "ready to attack the human species," but it seemed actually "to prefer preying on the human race rather than on any other animals."[77] Tigers were deemed not to be alone in this predilection. They shared it with several other contemptible animals, including wolves, who were characterized as "noxious," "savage," and "cruel" (also as afflicted with bad breath),[78] and the "ferocious," "insatiable," and "uncouth" polar bear.[79] Not so dangerous, but equally presumptuous, jackals and hyenas scavenged for human corpses.[80] But in a way the message was the same. Dead or alive, human flesh was forbidden fruit. These creatures were supposed to serve man's purposes, not appropriate him to theirs. To reverse this relationship was to rebel against the divine order, to commit sacrilege.

The writers of natural history books for children in the late eighteenth and early nineteenth centuries liked to dwell on man-eating. It loomed far larger in texts than its frequency as a behavior among those species really capable of it or its likelihood as a fate for members of their audience would have justified. But if reading about the animal kingdom was also a way for children to learn how their own society was organized, then man-eating offered a serious lesson as well as an armchair thrill. It provided a graphic and extreme illustration of the consequences that might follow any weakening of the social hierarchy, any diminution of respect and obedience on one side and of firmness and authority on the other.

This kind of juvenile natural history, in which the animals, presented one by one, provided an implicit commentary on human social norms, was frequently reformulated and republished until the middle of the nineteenth century. And it did not vanish completely even then. Occasionally, subsequent scientific description of animals served a double function by instructing children about the rules governing the human world. Between the lines of Arabella B. Buckley's *Winners in Life's Race*, for example, lurked the sternest social Darwinism, although she attempted to mitigate it by declaring that "the struggle is not entirely one of cruelty or ferocity, but . . . the higher

Learning from Animals

"The Tiger," from James Rennie, *The Menageries*, 1829.

the animal life becomes, the more important is family love and the sense of affection for others."[81]

On the whole, however, moralizing dropped out of juvenile natural history literature in the middle of the nineteenth century. As science became more sophisticated, the very term *natural history*, which had an aura of amateurism and speculation, gave way to soberer, more precise rubrics. Buckley's book itself exemplified this trend. The introduction and conclusion provided a didactic context for a text that was otherwise stuffed with Latinate taxonomical terms, paleontological evidence, and an unremitting concern with adaptation to function. As the title of one of Buckley's other works, *The Fairyland of Science,* suggests, she wished to introduce children to accurate zoological ideas. The moral dimension was a kind of sugar coating, not an integral part of her demonstration.

As well as changing the tone of juvenile nonfiction about animals, the Victorian advance of science undermined the metaphor equating subordinate human groups with animals in a more profound way. If Darwinian evolution were acknowledged, man had to be included among the animals; the once-impassable gulf of reason ceased to matter. Although Buckley did not go so far as to treat man in her survey of vertebrates (organized by functional and developmental groups, rather than creature by creature), she did include him in her "Birds-eye View of the Rise and Progress of Backboned Life" as "the last and greatest winner in life's race." He was not intrinsically separate from "large wild animals" but was their competitor for "possession of the earth."[82]

In earlier natural history literature for children, the metaphorical equation of inferior humans and inferior animals derived much of its appeal from the implicit assumption that the human social world was somehow nicer as well as more civilized than that of even domestic animals. Understood in the context of an unbridgeable gap between quadrupeds, the similarities between animals and people made it possible to teach children lessons about hierarchy and power that might have been unpleasant, even frightening, if expressed directly. As zoology brought animals and people closer together, real animals became inappropriate carriers of moral lessons. Only animals that had been humanized and sentimentalized could be admitted into Victorian nurseries as teachers. Learning about themselves from animals became the exclusive prerogative of readers of the other, fictional branch of animal literature for children, where it continued to flourish, producing such sentimental favorites as Black Beauty, Toad of Toad Hall, and the Cowardly Lion.

Notes

1. Darton, *Children's Books in England,* 1; Plumb, "First Flourishing of Children's Books," xviii–xix, xxiv. For a general discussion of edifying children's literature in the eighteenth century, see Pickering, *John Locke and Children's Books.*

2. Freeman, "Children's Natural History Books before Queen Victoria," 8, 11. The second installment of Freeman's article, subtitled "A Handlist of Texts," appeared in the autumn of 1976. See also Quayle, *Collector's Book of Children's Books,* 28.

3. See David Ellison Allen, *Naturalist in Britain,* chaps. 2 and 3, for a discussion of how natural history became a fashionable pastime. Keith Thomas's *Man and the Natural World* includes a survey of attitudes toward nature in early modern Britain. For a survey of the history of Western attitudes toward animals, see Passmore, "Treatment of Animals."

4. Darton, *Children's Books in England,* 10–12; Plumb, "First Flourishing of Children's Books"; *Fables of Aesop, and others,* iii. Eugene Francis Provenzo offers a history of the fable, with special attention to its didactic function, in "Education and the Aesopic Tradition," chap. 2.

5. James, *Bestiary,* 2–3, 7, 22; Elliot, *Medieval Bestiary,* n.p.

6. Topsell, *Historie of Foure-Footed Beastes,* title page.

7. Boreman, *Description of Three Hundred Animals,* preface, n.p.

8. Teltruth, *Natural History of Four-footed Beasts,* 11, 62. Modern bibliographers are usually generous in their classification, including as juvenile literature any work that might conceivably have attracted children. See Freeman, "Children's Natural History Books," 9–10, and his practice in the second installment, "A Handlist of Texts."

9. *Pretty Book of Pictures,* 21.

10. Boreman, *Description of Three Hundred Animals,* 6, 22, 19, 27. Boreman was not alone in either his credulity or his skepticism. Topsell had doubted the existence of the unicorn more than a century earlier (*Historie of Foure-Footed Beastes,* 711–21); on the other hand, Teltruth included it in *The Natural History of Four-footed Beasts.*

11. Boreman, *Description of Three Hundred Animals,* 67.

12. Bewick, *General History of Quadrupeds* (1822), 178.

13. Examples of this genre include Bennett, *Tower Menagerie;* and Frederica Graham, *Visits to the Zoological Gardens.*

14. Pickering, *John Locke and Children's Books,* 70–71; Plumb, "First Flourishing of Children's Books," xvii–xviii.

15. Boreman, *Description of Three Hundred Animals,* preface, n.p.

16. [Jones], *Natural History of Beasts,* iii.

17. Darton, *Children's Books in England,* 5; Plumb, "First Flourishing of Children's Books," xviii; Kramnick, "Children's Literature and Bourgeois Ideology," 211–12. Sylvia Patterson dissents from the scholarly consensus that identifies the middle classes as the primary producers and consumers of children's literature in the eighteenth century, claiming that upper-class children were the primary targets. Patterson, "Eighteenth-Century Children's Literature in England," 38–39.

18. Rennie, *Alphabet of Zoology,* 5–6.

19. [Jones], *Natural History of Beasts,* ix.

20. Mary Trimmer, *Natural History,* 4.

21. Holloway and Branch, *British Museum,* 1:iii.

22. *Natural History of Domestic Animals,* vi.

23. Buckley, *Winners in Life's Race,* viii.

24. See, for example, Sarah Kirby Trimmer, *Fabulous Histories;* and Pratt, *Pity's Gift.*

25. [Fenn], *Rational Dame,* vi.

26. Ibid., 19, 22.

27. *Animal Museum,* iii–iv.

28. Varty, *Graphic Illustrations,* n.p.

29. *Natural History of Domestic Animals,* v. 92; *Pretty Book of Pictures,* 30.

30. Holloway and Branch, *British Museum,* 2:45–48.

31. Mary Trimmer, *Natural History,* 4.

32. *Animal Sagacity,* 3, 5, 8.

33. Mary Trimmer, *Natural History,* 4, 9.

34. Pickering, *John Locke and Children's Books*, 25–33; Darton, *Children's Books in England*, 3.

35. *Animal Museum*, 204–7. See also [Jones], *Natural History of Beasts*, 65; and *Natural History of Animals*, 46–47.

36. [Jones], *Natural History of Beasts*, 73; *Tom Trip's Museum*, pt. 2, p. 11; Boreman, *Description of Three Hundred Animals*, 26; William Bingley, *Animal Biography*, 1:36.

37. Teltruth, *Natural History of Four-footed Beasts*, 72–73.

38. *Animal Museum*, 1.

39. [Jones], *Natural History of Beasts*, ix.

40. *Natural History of Domestic Animals*, 84, 106.

41. *Animal Sagacity*, 28–30, 130–32.

42. Holloway and Branch, *British Museum*, 2:181.

43. *Tom Trip's Museum*, pt. 1, p. 6; Boreman, *Description of Three Hundred Animals*, 11; Thomas Bingley, *Stories Illustrative of the Instincts of Animals*, 65.

44. *Tom Trip's Museum*, pt. 1, p. 3; *Animal Museum*, 1.

45. Holloway and Branch, *British Museum*, 1:145.

46. *Tom Trip's Museum*, pt. 1, p. 3.

47. Thomas Bingley, *Stories Illustrative of the Instincts of Animals*, 15; *Animal Museum*, 1.

48. [Jones], *Natural History of Beasts*, 7.

49. Ibid., 79.

50. *Animal Sagacity*, 55–56, 97–98.

51. *Natural History of Domestic Animals*, 9; [Fenn], *Rational Dame*, 41.

52. [Fenn], *Rational Dame*, 41.

53. Holloway and Branch, *British Museum*, 1:31.

54. *Natural History of Domestic Animals*, 78–82.

55. [Fenn], *Rational Dame*, 36.

56. [Jones], *Natural History of Beasts*, 50.

57. Mary Trimmer, *Natural History*, 25–26; [Fenn], *Rational Dame*, 38.

58. [Jones], *Natural History of Beasts*, 93; *Natural History of Domestic Animals*, 64–67; [Fenn], *Rational Dame*, 23.

59. *Animal Museum*, 162.

60. [Jones], *Natural History of Beasts*, 62; *Natural History of Animals*, 13; Holloway and Branch, *British Museum*, 44; Mary Trimmer, *Natural History*, 39.

61. [Fenn], *Rational Dame*, 53.

62. Mary Trimmer, *Natural History*, 81; Holloway and Branch, *British Museum*, 2:136.

63. *Tom Trip's Museum*, pt. 3, p. 13; [Jones], *Natural History of Beasts*, 55.

64. *Natural History of Animals*, 53.

65. Mary Trimmer, *Natural History*, 41; [Jones], *Natural History of Beasts*, 34, 59, 115; *Pretty Book of Pictures*, 30–31.

66. [Jones], *Natural History of Beasts*, xi.

67. Ibid., 117.

68. *Animal Museum*, 93.

69. *Tom Trip's Museum*, pt. 2, p. 2; *Animal Museum*, 168–71.

70. Park, *Travels in the Interior of Africa*, 113.

71. William Bingley, *Animal Biography*, 1:268.

72. Mary Trimmer, *Natural History*, 17.

73. Holloway and Branch, *British Museum*, 1:29.

74. *Animal Museum*, 173.

75. [Jones], *Natural History of Beasts*, 106; *Tom Trip's Museum*, pt. 2, p. 8; Holloway and Branch, *British Museum*, 2:25.

76. Holloway and Branch, *British Museum*, 1:22.

77. *Natural History of Animals*, 11; [Jones], *Natural History of Beasts*, 101.

78. *Natural History of Animals*, 35; [Fenn], *Rational Dame*, 44; [Jones], *Natural History of Beasts*, 86; Holloway and Branch, *British Museum*, 1:54.

79. Mary Trimmer, *Natural History*, 43; Shoberl, *Natural History of Quadrupeds*, 2:169; Holloway and Branch, *British Museum*, 1:222.

80. Holloway and Branch, *British Museum*, 2:245; Shoberl, *Natural History of Quadrupeds*, 2:72–73.

81. Buckley, *Winners in Life's Race*, 351–52.

82. Ibid., 343–45.

— 3 —

Toward a More Peaceable Kingdom

Supplicants of various sorts tend to congregate at the entrance to my local supermarket. Over the years, I have watched my fellow shoppers firmly ignore candidates for state and local office, solicitors for worthy charities, and advocates of principled positions on important ethical issues. Once, however, I was startled to see a long line of people patiently waiting to sign a petition. The inspiration for this unprecedented show of enthusiasm was the campaign led by the Massachusetts Society for the Prevention of Cruelty to Animals (MSPCA) against the state's pound seizure law, which allowed research institutions to appropriate pets unfortunate enough to end up in shelters. The organizers were trying to collect the number of signatures necessary to place a referendum question on the next statewide ballot. They were successful in their petition drive, and ultimately, despite a counter-campaign mounted by the biomedical establishment, they proved able to attract votes as effectively as they had attracted signatures. In 1983 it became illegal to recycle former pets for research in Massachusetts laboratories.

Although my home state has often enjoyed a reputation for political eccentricity, the attitude of its citizens toward pound seizure puts them firmly in the American mainstream. More than ten other states have similar legislation, and in many of the rest, prohibition of the experimental use of shelter animals is a widely exercised local option. Protest against pound seizure is only one manifestation—and a rather mild one—of antivivisectionism, or re-

---

"Toward a More Peaceable Kingdom" originally appeared in *Technology Review* 95, no. 3 (February/March 1992): 55–61.

sistance to the use of living animals in scientific experiments. The public is easily moved to shock and pity when the potential experimental subjects are dogs and cats, our most common and best-loved pet species, especially when the individual animals in question have been socialized to regard humans as friends.

Moreover, advocates of pound seizure have been unable to argue persuasively that its prohibition would interfere with vital research, and in fact, the relatively few scientists who might have requisitioned pound animals have adjusted to buying what are referred to as "purpose-bred animals"—that is, animals produced in breeding laboratories for lives and deaths as experimental research subjects. Thus, despite its broad appeal and the misgivings it aroused in the research community, where any external attempt to influence procedures or decisions can be cause for alarm, the pound seizure controversy faded quickly from public consciousness.

But the sympathy for animal suffering that the controversy exposed did not subside. On the contrary, it has been, if anything, exacerbated by the more radical antivivisection issues that have replaced pound seizure in the headlines. These issues directly engage fundamental oppositions, pitting scientific values against more general moral commitments, human needs and interests against those of other animals, and relatively analytic ways of thinking against more holistic ones.

The growing concern for the suffering of animals in laboratories has had far-reaching practical consequences as well. A sustained media attack on toxicological testing has contributed to the success of enterprises like the Body Shop, a cosmetics chain that distributes free antivivisection literature as it sells shampoo. Such pressure has also led several major cosmetics manufacturers to stop using animals to test the safety of their products.

The use of live animals in biomedical education has been subjected to court challenges by students at every level, from secondary school to veterinary school. And specific research agendas—particularly those in which experimenters inflict pain or trauma on fellow primates—have been the targets of protests ranging from picketing to burglary and arson. Radical activist organizations have been nearly as successful at recruiting members as the moderate MSPCA was at collecting signatures. For instance, People for the Ethical Treatment of Animals, or PETA, currently one of the most prominent and controversial animal advocacy groups, could count only eight thousand members in 1985, but the figure swelled to a quarter of a million over the

STUPIDITY AND SCIENCE.
(Meeting of Medical Professors.)

Punch cartoon, 1876.

next five years. Meanwhile, many researchers have organized to resist what they regard as an assault on both their work and the scientific institutions necessary to sustain it.

As this escalating activity suggests, the antivivisection movement addresses concerns of special contemporary relevance. And passions have been running high on both sides. But if the heat of battle signals the intensity of current engagement with these issues, it may also obscure the underlying framework of the debate.

Can They Suffer?

The modern antivivisection movement is often dated from the 1975 publication of Peter Singer's *Animal Liberation*. In this landmark combination of philosophy and common sense, Singer developed a powerful "utilitarian" argument for treating the interests of animals as morally significant. That is, he maintained that the pleasure or pain experienced by animals had to be taken into account when calculating the moral value of a given action, and, further, that any human pleasure or pain did not automatically outweigh any animal pleasure or pain. He went on from there to offer instructions to the converted on how to change their lives. (These instructions were rather detailed for a philosophical work, including hints about vegetarian cooking.)

Animal Liberation extended the egalitarian thrust of the protest movements of the 1960s; antivivisectionism, along with related campaigns such as those against factory farming and the exploitation of performing animals, placed animals under the umbrella of concern then newly enlarged, at least in principle, to include humans of all races and both genders. And even if the generous spirit of the 1960s has recently seemed to be in decline or even in eclipse, radical animal activism has offered a bridge connecting it with succeeding zeitgeists. Thus it is easy to associate animal advocacy with vegetarianism, environmentalism, and multiculturalism.

Modern and even modish though it may be, however, antivivisectionism is not new. It is almost as old as the practice of using live animals as experimental subjects, which itself is at least as old as modern experimental science. Even in those early days, vivisection could seem to typify scientific practice in general. For example, Francis Bacon, who died in 1626 of bronchitis, is frequently cited as a martyr to science, but he was actually a martyr to vivisection; he caught cold while collecting snow with which he planned to refrigerate a chicken.

The emergence of lay criticism of animal experimentation is associated with the seventeenth- and eighteenth-century increase in scientific research of all kinds. Interest in discoveries about the natural world extended far beyond the ranks of serious investigators, and during the first part of the eighteenth century there emerged a thriving market for popular science, including books, instruments, specimens, and scientific demonstrations (these entertainments probably filled more or less the same niche that television programs like NOVA occupy today). Armed with the self-serving assurance that animals were really unfeeling automata, scientific impresarios cut open howling dogs to display their internal organs and asphyxiated small birds with air pumps. Not surprisingly, some observers were repelled by such spectacles and unpersuaded by the assurances; occasionally one of these nonbelievers managed to administer impromptu euthanasia when the demonstrator's attention was diverted.

Experimentation on live animals also began to attract serious religious and philosophical criticism during this period. In his *Introduction to the Prin-*

"Am Not I a Brute and a Brother?" from *Punch*, 1869.

ciples of Morals and Legislation the English philosopher Jeremy Bentham—a utilitarian and therefore the intellectual ancestor of Peter Singer—offered the best-known objection. He asserted that "the question is not, Can they reason? Nor, Can they talk? But, Can they suffer?"

In 1789, however, vivisection was relatively low on the list of human practices that animals had to fear. Pioneering animal protection legislation enacted in Great Britain in the 1820s and 1830s focused on the abuse of animals used in commercial pursuits—for example, the overloading of cab horses—and on such lower-class blood sports as bullbaiting and dogfighting. Nevertheless, the issue of antivivisection was raised during the formative period of the Society for the Prevention of Cruelty to Animals,[1] which was founded in 1824 to give teeth to the new laws. The authors of the SPCA's initial prospectus outlined the equivocal position that the mainstream humane movement has often reendorsed since: they deplored anything that could be characterized as mere cruelty in the name of science, but admitted that under responsible and benevolent direction vivisection might be justifiable.

They were not required to match actions to these words—or even to elaborate upon them—for another generation. During the first half of the nineteenth century, vivisection became a common practice in the laboratories of continental Europe, where such disciplines as physiology and immunology were emerging, but British researchers tended to prefer less invasive approaches. Although this national investigative tradition was to lead to one of the most important scientific achievements of the nineteenth century, Darwin's theory of evolution by natural selection, it ultimately became clear that if British scientists were to participate in vanguard biological research, they would have to adopt Continental methods. And as vivisection became more common, objections began to arise. Some resistance came from within the scientific and medical communities—usually from old-fashioned practitioners who could foresee that their own standing would be undermined by the new techniques—but most came from members of the public who were simply horrified by what they heard of the sufferings undergone by experimental subjects.

The heated and protracted national debate that ensued had far-reaching consequences. It split the animal protection movement into moderate humanitarians—pragmatists who were willing to balance human and animal interests—and radical antivivisectionists, who were ideologically opposed to any kind of experimentation on animals. Conversely, it unified the scientific community and inspired what may have been the first organized campaign of scientists to protect their own political interests. Even the elderly Darwin, who had long since abandoned the field sports of his youth and who would not have dreamed of experimenting on live animals himself, enlisted on the side of vivisection, having been persuaded that the future of science itself was at stake.

The controversy culminated in the Cruelty to Animals Act of 1876, the provisions of which were stronger than those of any comparable national legislation that has yet been enacted in the United States. However, instead of prohibiting animal experimentation, the act merely established regulatory procedures, so it was generally perceived as a triumph for the scientific establishment and a defeat for organized antivivisectionism.

The Man in the White Coat

Interestingly, that defeat extended beyond the official sphere into the court of public opinion: gradually, over the next few decades, the antivivisection movement lost its broad-based following and was marooned at the political periphery, where it stayed until its recent renascence. Even more interestingly, the concerns that underlie antivivisectionism have changed remarkably little in the last one hundred years, which may complicate our sense of its relationship to other social issues.

For example, the analogy between animals and oppressed human groups that has appealed so strongly to modern animal advocates was less broadly resonant in the late nineteenth century, but it was not absent. Then as now, animal protection organizations of all types had a special appeal for women, perhaps because women were inclined to protect creatures whose helplessness reminded them of their own social position. The advent of gynecology in particular, with its intrusive metal instruments and enforced supine immobility, inspired some nineteenth-century women to identify concretely with animals laid out on the operating table. Members of the working classes also occasionally expressed antivivisectionist feeling, probably reflecting their not unreasonable fear that humble human corpses, however decently buried, would end up under the dissecting knives of anatomists and medical students.

But a much stronger connection between the antivivisection movement of a century ago and that of today emerges if our focus shifts from the image of the passive experimental subject to that of the active researcher. The white-coated laboratory scientist is a familiar symbol of the disciplined and unbiased search for truth and, in the specifically biomedical context, of the search for new knowledge that will alleviate human suffering. From this perspective the researcher seems at once authoritative, upright, and beneficent.

There has long existed, however, a counterinterpretation of this image, in which the immaculate figure appears arrogant and irresponsible, overbearing and cruel. From this perspective the researcher represents an institution devoted not to truth but to self-aggrandizement at the expense of other, often more traditional and less elitist sources of expertise—an institution that follows its own rules, even when these contravene the moral consensus that guides the rest of society. It is this counterinterpretation of the scientist and the institution of science that most powerfully links the antivivisectionists of different eras.

Thus the popular appeal of antivivisectionism tends to rise in direct proportion to public skepticism about the value of scientific research and public resentment of the behavior of scientific researchers. The nineteenth-century peak in antivivisectionist activity reflected widespread shock at the methods of the new biological science, which violated conventional religious attitudes at least as strongly as Darwin's theory of evolution by natural selection did. And public appreciation of such practical benefits of physiological research as insulin and the diphtheria antitoxin explains, at least in part, the turn-of-the-century decline in protests. This shift in opinion initiated a long halcyon period in the relationship between the public and biomedical science. Scientists and laity agreed, for the most part, that the white-coated figure was owed respect and deference, that his (the pronoun is used intentionally) authority over his own domain should not be questioned, nor his research methods and goals second-guessed.

It is always difficult to date unbounded occurrences such as shifts in general opinion, but sometime after the Second World War this view of the research enterprise began to seem less compelling, at least outside the scientific community. The most tangible scientific advances apparently led not to healing but to destruction—most spectacularly, of course, the atomic bomb, but also, looking backward, the chemical warfare of the First World War. Moreover, despite generous and sustained government financial support, no scientific fixes have been forthcoming for the terrible degenerative diseases such as cancer and cardiovascular illness that have killed most Americans since the nineteenth-century triumphs of immunology and public health. And, at best, science has played an ambiguous role in the snowballing environmental crisis. As the various sources of skepticism about the scientific enterprise converged, the time was ripe for the reemergence of antivivisectionism. Even though both the world and biomedical science had changed significantly, the concerns that inspire animal activism had scarcely altered.

Indeed, in 1896—that is, in the twilight of the earlier antivivisection movement—H. G. Wells published an underappreciated novel, *The Island of Dr. Moreau*, which shows how relevant the issues of a century ago still seem. Wells's book chronicles the downfall of an ambitious researcher who, banned from Europe because of his outrageous experimental practices, has retreated to a remote South Sea island to continue his work, which turns out to be the construction of human beings from other animals. The narrator, an English dilettante inadvertently stranded in Dr. Moreau's domain, stresses both the painfulness of the procedures and the presumptuousness of the goal; the island is like a nightmare version of the Garden of Eden.

A modern reader is apt to be struck first by the quaintness of Wells's science—when I taught *The Island of Dr. Moreau* to undergraduates at MIT, my students quickly pointed out the enormous rejection problems that would

The Animal World, vol. 7, 1876.

be produced by cobbling together organs extracted from animals of different orders—but it soon becomes clear that Wells's underlying critique of experimental science needs very little alteration to bring it up to date. (Even the science can seem timely enough, if allowances are made. For example, if a quest to create new life forms through the surgical manipulation of animals has become inconceivable, genetic engineering presents a range of analogous issues.)

Wells portrays Moreau's obsessive pursuit of knowledge as both inhuman and inhumane. The researcher inflicts pain on his animals only to gratify his own curiosity and ambition; the same motives lead him not only to ignore the structures of society, but also to remove himself from it physically. He contemptuously dismisses the notion that there could be any countervailing ethical considerations powerful enough to persuade him to forgo the acquisition of new knowledge.

It should be emphasized, too, that Wells's censure is independent of Moreau's scientific success. In other words, Moreau's presumption is punished and he ultimately perishes with all his work, even though he actually produces humanoid creatures who can talk, think, and walk on two feet.

The Pursuit of Compromise

Wells's critique of 1896 seems current for much the same reason that Bentham's statement of 1789 continues to be quoted. The underlying objections to vivisection derive from unchanging convictions about the proper relation of humanity to nature, and the practice can come to represent a range of human intrusions into areas that, from a moderate perspective, would be better left alone, and, from a radical one, are despoiled by this contact.

Unfortunately, it is at least difficult, and usually impossible, to reconcile such antivivisectionist convictions with the aggressively inquisitive ideology that underlies commitment to scientific research. This difficulty has been implicitly recognized in the current controversy, most tellingly by the consistent reluctance of the embattled parties to engage each other seriously. There has been a lot of name-calling—the bloody epithets hurled at scientists have been countered by assertions, sometimes angry, but more often blandly condescending, that anyone with antivivisectionist sympathies is either crazy or ignorant. Even when these assertions are not openly made, they are frequently implied. For instance, an article in *Issues in Science and Technology* by Jerod Loeb of the American Medical Association and Deborah Runkle of the American Association for the Advancement of Science suggests that ordinary citizens could be induced to abandon any antivivisectionist inclinations they might harbor if only experimenters would take a little time away from their research and devote it to sharing their expert opinions with the public. What is disparaging about this proposal is that the communication is envisioned as a one-way street, as if the public mind were a vacant and inert receptacle, requiring only to be filled with the right ideas.

One reason rapprochement has been so elusive is that in the vivisection debate—as in the debates about abortion and euthanasia—biology, technology, and ethics intertwine, so that the kind of give and take that leads to pragmatic compromise is particularly hard to come by. Given the basic premises of each side, the only logically consistent positions turn out to be the impractical and inflammatory extremes: either scientists can do anything they want with animals or they can do nothing at all. As a result, people who seek

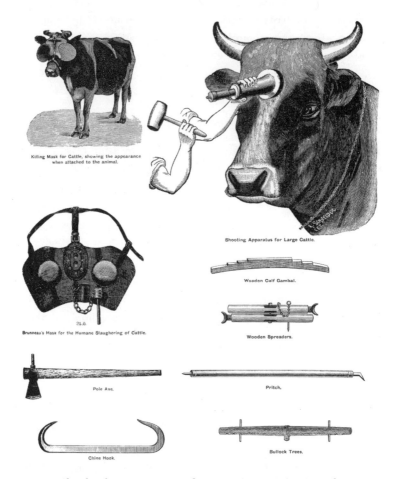

Slaughterhouse equipment, from *Douglas's Encyclopedia,* n.d.

a middle way just end up bedeviled by questions of degree. Why do some research agendas warrant more animal suffering than others? Why should primates receive one kind of treatment and cats another? What is it about rats and rabbits that qualifies them for less humane treatment? What about frogs and fish?

Nevertheless, there has been some progress. Many parties to the vivisection controversy have recognized that when the logical comes into conflict with the pragmatic, the better part of valor may be to abjure absolute consistency as an unaffordable luxury and settle for reasonableness. Accordingly, they have tried to balance the welfare of individual animals against the social and scientific good that might result from some research. The mainstream humane movement has generally supported limited animal experimentation under humane conditions and for worthwhile purposes. Thus in 1988, when *Animals,* the magazine of the MSPCA, reported sympathetically on the ethologist Jane Goodall's campaign for better care of chimpanzees used in AIDS research, one of the main points the author made was that improved conditions would produce more reliable experimental results. Recently a few scientific organizations, such as the Scientists' Center for Alternatives to Animal Testing, have been founded to pursue goals similar to those of the MSPCA, and, albeit too slowly, government regulation has also been moving in this direction.

But the real obstacle to constructive change may be the power dynamics that underlie the antivivisection debate. As long as those dynamics persist, efforts from either side are apt to be undermined by the logical appeal and loud voices on their extreme flank. Significantly, the traditional disruptive tactics of antivivisectionism—protests and demonstrations, the marshaling of public outrage—have been deployments of the power of the weak. And however threatening and troublesome the antivivisection movement has occasionally seemed, it has always been weak in relation to biomedical science; that is, it has never had much impact on what actually happens to animals in laboratories.

A lasting resolution of this recurrent conflict would require some genuine sharing of the power of the strong, a restructuring of institutions and procedures to open them to lay scrutiny and influence (real influence, not the token representation that characterizes many university committees on animal care and use). Yet, as has been repeatedly demonstrated in arenas far removed from laboratory science, the strong are not easily moved to such generosity. Antivivisectionists are not the only committed ideologues who have occasionally been pushed to acts of terrorism.

Indeed, the difficulty of accomplishing the necessary reforms would be hard to overestimate, partly because they would inevitably lead to sizable reductions in the number and kind of experimental procedures performed on animals, thus disrupting the research programs of many scientists. Another roadblock is that scientists, like most other groups defined by expertise, are strongly inclined to defend their professional turf and resist external attempts at regulation or governance. Most problematic of all, though, the ideology of science itself will be challenged, forcing experimenters to recognize that they are not necessarily carrying out an independent exercise in the pursuit of truth—that their enterprise, in its intellectual as well as its social and financial dimensions, is circumscribed and defined by the culture of which it is an integral part. This ideology is deeply entrenched and upheld with passionate commitment. But until some minds change, the marching will continue.

Notes

This essay originally appeared in 1992. The account of recent events was based on information provided by news media, government agencies, and other interested organizations, both in print and electronically. For more extended accounts of the history of animal protection and antivivisection, see Richard D. French, *Antivivisection and Medical Science;* Guerrini, *Experimenting with Humans and Animals;* Kean, *Animal Rights;* Lansbury, *Old Brown Dog;* Ritvo, *Animal Estate;* and Rupke, *Vivisection in Historical Perspective.*

1. It became the Royal Society for the Prevention of Cruelty to Animals in 1840.

— 4 —

Animal Consciousness

Some Historical Perspective

Historical background is often understood to mean a survey of the past, in order to show how it prepared the way for the present state of the field (whatever field may happen to be under discussion)—in other words, a kind of intellectual genealogy or search for ancestors. And such an overview is a fine thing to have, certainly, for various reasons. It illustrates the deep historical roots and resonances of even the most up-to-the-minute research, an aspect of their work that investigators focused on novelty and originality are apt to overlook. Further, an appreciation of the difficulties posed to able investigators of earlier periods by problems whose solutions retrospectively seem simple or obvious or even beside the point may encourage us to reflect on the way we have formulated and assessed the challenges that we face now. Finally, such straightforward recapitulation allows the apportionment of credit where credit is due. There is no question that we all (even historians, who tend to be less strongly committed than scientists to a progressive understanding of our discipline) stand on the shoulders of our predecessors.

But, as is the case with most attempts to make sense of large and disparate bodies of information, the illumination that such a retrospective account provides inevitably has its (literally) dark side. That is, by drawing our attention to the activities and accomplishments of some predecessors, it diverts our attention from others. Think, for example, of the metaphor I have just invoked

---

"Animal Consciousness: Some Historical Perspective" originally appeared in *Integrative and Comparative Biology* 40, no. 6 (December 2000): 847–52.

(not my own, I hasten to add)—that is, of current investigators standing on the shoulders of those who have gone before them. This metaphor compares intellectual progress either to a kind of acrobatic performance or to the way that parents sometimes hoist their children to give them a clearer view. It rightly emphasizes the long (often ancient) pedigree of even the newest insights. But at the same time it offers several misleading suggestions—most forcefully, that there is a single line of descent that leads from the past cutting edge to the present cutting edge (apologies for that mixed metaphor), but also, for example, that the work that seems most attractive to us as we look back was most highly esteemed by contemporaries, and that, even where this was the case, it was appreciated for the same qualities or characteristics that make it appeal to us. In sum, it suggests that the past can be understood simply as the forerunner of the present. In this suggestion, it is coherent with most of the other metaphors that condition our understanding of the intellectual history of our current pursuits—for example the metaphor of genealogy or descent that I have already mentioned. But now I would like to suggest a somewhat cross-grained extension of this metaphor: that is, to put it in a nutshell, it is a wise discipline that can recognize its own forebears.

Let me state this point in a different way: the reductiveness of the usual kind of "historical background" has its appeal, and also its utility. "Potted history" (another term for the same thing) tends to be easily consumable, both because it is brief and uncomplicated, and also, often, because it flatters both its intended audience and the preconceptions cherished by that audience. But it also can be significantly misleading—and not just about the past. The monodimensional understanding retroactively imposed on the past tends to characterize our understanding of the present as well. For example, current debates about animal consciousness, among others, are also conditioned by the assumption that the stakes are absolute. Positions are understood to be either right or wrong, and the consequence of their rightness or wrongness is understood to be either intellectual survival or intellectual extinction. If we look back, however, we see that, at least in the past, such clarity has not been the rule. I would like to propose a somewhat expanded notion of historical background—one that points to the variety of past understandings of animal consciousness, the diversity of causes that made people embrace or reject them, and the difficulty of deciding which ones to privilege as our precursors.

Of course, as soon as we begin to look at the past of any modern field of endeavor we encounter a basic problem of historical evidence and interpretation—that of language or definition. Both of the terms in the title of this essay—*animal* and *consciousness*—are contentious at present. They cannot be defined without reference to ideology. And they have been equally and analogously contentious in the past. At issue—whether explicitly or implicitly, whether in philosophical or religious terms—is the position of human beings within the order of creation. Can we stand to see ourselves as less than singular, and, if not, why not? Aside from this weighty complication, there are more ordinary difficulties—difficulties that complicate the work of understanding all precontemporary texts. A century or two is time enough at least to alter the connotation of most words, even if they retain their form and their ostensible reference. And often enough, they do not retain their ostensible reference. That is, they may shift in denotation—in literal meaning—as well as in connotation; or they may be replaced by completely different terms. The words in the essay title exemplify both a subtle shift and replacement.

The main current problem with the term *animal* is taxonomic: does it include people, and if so, are people included in only their physical aspect or in some more encompassing sense? This problem is embedded in deep etymology, since *animal* descends from the Latin word *anima*, the varied meanings of which include "breath," "life," "soul," and "mind." In the eighteenth and nineteenth centuries the line between people and other animals was both more and less starkly drawn than it is at present. Both convictions and uncertainties were expressed through classification. Enlightenment naturalists often recognized not only the general correspondence between people and what were then known as quadrupeds, but also the more striking similarities that human beings shared with apes and monkeys.

For example, in his 1699 treatise on the chimpanzee (which he called the "orangutan" or "pygmy") the anatomist Edward Tyson implicitly included humanity in the animal series.[1] His choice of terminology further implied that the categories "human" and "orangutan" might not be completely distinct. His Latin term for the creature was *"Homo Sylvestris"* or "wild man of the woods," and, conversely, the humanity of the quasi-mythical pygmies had long been the subject of European speculation. Even at the end of the eighteenth century, naturalists could claim that pygmies were "nothing more than a species of apes . . . that resemble us but very imperfectly."[2] (When thinking about the boundaries of humanity in earlier periods it is also well to remember that until relatively recently opinion was sharply divided about whether everyone now included in *Homo sapiens* was fully human.)

The portrayal of apes as particularly human in appearance and behavior

Frontispiece to Edward Tyson's *Orang-outang*, 1699.

extended this implicit assault on the human-animal boundary. Illustrations in works of natural history frequently showed apes assuming erect posture, using human tools, and approximating human proportions in the trunk and limbs. The chimpanzees and orangutans who were predictable features of nineteenth-century zoos and menageries ate with table utensils, sipped tea from cups, and slept under blankets. An orangutan who lived in London's Exeter Change Menagerie amused herself by carefully turning the pages of an illustrated book. At the Regent's Park Zoo a chimpanzee named Jenny regularly appeared in a flannel nightgown and robe. Consul, a young chimpanzee who lived in Manchester's Belle Vue Zoological Gardens at the end of the nineteenth century, dressed in a jacket and straw hat, smoked cigarettes, and drank liquor from a glass.[3]

Animal Consciousness

If the term "animal" has been the subject of subtle shifts, "consciousness" is the most recent in a series of replacements. In the nineteenth century, for example, most discussion of the mental qualities of animals took place under the rubrics of "intelligence" and "sagacity." Then as now, such discussions were not merely descriptive; after all, intelligence and sagacity are no easier to pin down than is consciousness. The quality or qualities under consideration, however denominated and defined, were the focus of particular interest because they contributed both to a ranking of nonhuman animals among themselves and to a consideration of how closely they approached the human condition. One way of denying the human-ape connection—as of denying the connection between human groups—was to posit an alternative alliance. If nonprimate animals resembled humans more closely than did apes, then they would necessarily displace apes from their awkward proximity.

Thus, throughout the nineteenth century naturalists debated the rival

claims of dogs and apes to be top animal, and therefore closest to humankind. In 1881, for example, George J. Romanes, a close friend of Darwin's with a special interest in animal behavior, celebrated the "high intelligence" and "gregarious instincts" of the dog, which, he argued, gave it a more "massive as well as more complex" psychology than was possessed by any member of the monkey family.[4] Two years later Romanes revised his ranking slightly, including both dogs and apes on level twenty-eight of his famous fifty-step ladder of intellectual development. Level twenty-eight was characterized by "indefinite morality" along with the capacity to experience shame, remorse, deceit, and the ludicrous. (To give some sense of the scale: steps twenty-nine through fifty were reserved for human beings, while worms and insect larva occupied step eighteen because they possessed primary instincts and could feel the emotions of surprise and fear.) Although this schema gave apes and dogs equivalent rank, Romanes was far from thinking that they possessed identical mental attributes. Rather the ape had achieved its high status through intellect, the dog on account of highly developed emotions.[5]

At issue was how to define animal intelligence—if, indeed, animals could be said to possess intelligence at all. Some nineteenth-century naturalists denied that animals possessed any mental qualities besides instincts. A correspondent of the *Zoological Journal* asserted that although dogs and other animals exhibited behavior that closely mimicked such qualities as foresight, industry, and justice, in fact they were merely performing reflex actions, in the manner of Descartes' animal machines.[6] Most investigators were more generous, however, allowing the higher animals a grab bag of intellectual and emotional attributes. One representative inventory included imagination, memory, homesickness, self-consciousness, joy, rage, terror, compassion, envy, cruelty, fidelity, and attachment.[7]

An index of the mix of mental qualities that naturalists valued in animals—and perhaps also of their desire to distinguish clearly between animal and human mental capacities—was the fact that well into the last part of the nineteenth century "sagacity" was the standard term for intelligence demonstrated by animals. An individual animal or species might be described as "intelligent," but the term "intelligence" itself was generally reserved for strictly human capacities. (Conversely, if "sagacity" was attributed to human beings, it often had an ironic or less than flattering connotation.) The phrase "animal sagacity" in the title of a book or article often signaled an abstract discussion of instinct or intellect, the kind of discussion that might conclude by appreciating the intelligence of apes. But in the more common usage of naturalists, "sagacity" indicated not the ability to manipulate mechanical contraptions or solve logical problems, but a more diffuse kind of mental power: the ability to adapt to human surroundings and to please people. A somewhat circular calculation made the most sagacious animals the best servants. So dogs might not only rival apes in the mental competition but surpass them—closest to their masters in mind as well as in domicile. And since the alternative closeness thus constructed was clearly figurative, the whole animal creation was thereby implicitly removed to a more comfortable distance.

Even if we leave questions of definition aside, we find another source of confusion in retrieving the history of a discipline. It is customary to speak in chronological generalizations—to say that people thought this way in the Enlightenment, and that way in the Victorian period. But such generalizations beg important questions. They assume consensus—almost always an unwarranted assumption—and therefore require the selection of one among competing voices. It is worth asking on what basis such selection is made—how can we tell in retrospect who best represents a bygone era? Or, to put it in a different way, how do we decide whose opinions are most worth recalling? For example, with regard to the taxonomic story I have been recounting about the physical and behavioral resemblances between people and other animals: not everybody was persuaded by Tyson, or even, later, by Darwin. Commitments that were explicitly or essentially theological made many naturalists reluctant to embed their own species within the system of animal connections. And if physical resemblances were undeniable, that made it more important to defend the less tangible ground of mentation or behavior. Despite Carolus Linnaeus's sanctified status as a systematizer, his inclusive primate order was frequently rejected. According to the late-eighteenth-century naturalist Thomas Pennant, "my vanity will not suffer me to rank mankind with *Apes, Monkies, Maucaucos, and Bats*"; a contemporary similarly asserted that "we may perhaps be pardoned for the repugnance we feel to place the monkey at the head of the brute creation, and thus to associate him . . . with man."[8]

As evolutionary theory suggested a more concrete and ineluctable connection, it provoked still more forceful resistance. After *On the Origin of Species* was published in 1859, for example, the geologist Adam Sedgwick, who had been one of Darwin's early scientific mentors at Cambridge, asserted that "we cannot speculate on man's position in the actual world of nature . . . while we keep his highest faculties out of our sight. Strip him of these faculties, and he becomes entirely bestial."[9] As Darwin sadly noted at the end of *The Descent of Man,* "The main conclusion arrived at in this work, namely that man is descended from some lowly-organised form, will, I regret to think, be highly distasteful to many persons."[10] Given this diversity of opinion within the community of experts it is difficult to identify any particular view as distinctively characteristic of the Enlightenment or of the nineteenth century.

This is even true of one of the most frequently cited such views. We are accustomed to seeing Descartes and his animal machines as icons of the Enlightenment, along with Newton and his gravitational apple. For example Ernst Mayr, who has no love for "Descartes's crass mechanism" and who notes with pleasure the many critics of the biological aspects of his work, credits him with "the spread of the mechanistic world picture," and asserts that his claims "that organisms are merely automata" have "created a millstone around the neck of biology, the effects of which have carried through to the end of the nineteenth century."[11] But it is difficult to know why this position should be regarded as either especially typical of its period or uniquely ancestral to modern attitudes. Although it gave rise to (or served as justification for) many spectacular practices that have lent themselves to reproduction as part of "the rise of experimental science," there was a great deal of demurral among Descartes' contemporaries, both specialists and members of the interested general public. (Even members of the general public who were willing and even eager to inflict pain on animals—enthusiasts of badger baiting, fox hunting, and other blood sports—did not think that their victims could not feel it; on the contrary they often reveled in the fact that they did.) Sometimes this resistance was expressed formally, in the ripostes offered by investigators with greater experience of actual animals.[12] Sometimes it was expressed informally, as when spectators at public demonstrations quietly put the animal subjects out of their misery. And although this assertion can be seen as at the foundation of subsequent use of animals in experimentation, not all experimenters have been persuaded by it—and it also led to or justified many procedures that modern researchers find both shocking and silly.

Indeed, this particular case illustrates yet another recurrent problem of historical retrieval. When we look back, it can be difficult to decide who were the experts. With regard to the formal (taxonomic) relations that I have been discussing, the experts are mostly recognizable in modern terms—they were self-styled as naturalists if not as biologists or zoologists. But with regard to the sensibility of animals, a much wider circle of people has always felt able to speak with authority. It is not clear why Descartes rather than Bentham or some other like-minded *philosophe* should be regarded as the ancestral authority on such matters. It may be salutary to consider the somewhat different potted history of ideas about animal sensation offered by people whose major commitment is to the protection of animals, rather than to the increase of zoological knowledge.

Perhaps the modern study of animal consciousness should trace itself as much to the protesters as to the experimenters. That is, at least, if ancestry is determined by shared content as well as by shared form. This brings us to another complicating determinant of how we read our past—one that perhaps could be characterized as social rather than political. Descartes may be the preferred progenitor because he was scientific in his methods, at least some of them. This reading recapitulates a protracted effort within the study of animal consciousness—as within most other disciplines that fell within the nineteenth-century rubric of natural history—to cast out the amateurs, the people who were accused of wasting time in journals and at meetings with unsystematically recorded anecdotes about the intellectual and moral capacities of their anthropomorphic dogs, cats, and horses. This effort was characterized in terms of physics envy—such people and practices interfered with the attempt of what ultimately became known as animal psychology to become properly "scientific." This explanation is more persuasive if we look at the sinners rather than their judges; from a late-twentieth-century perspective they can be hard to distinguish. For example, although Romanes repeatedly warned of the pitfalls of anecdotal observation, he used his favorite terrier to illustrated the "exalted level to which sympathy had attained" and the "intelligent affection from which it springs" in the dog.[13] Conwy Lloyd Morgan described his observations of his own fox terrier as well as of several dogs belonging to friends (not to speak of a young chicken named Blackie) as experiments.[14] And, in any case, the kind of information that had been excluded as anecdotalism reemerged in a more respectable form within only a few decades as ethology. Research about animal consciousness still comes from a range of disciplinary and subdisciplinary sources. It may be worth asking whether similar questions of sociology still influence relations between exponents of these various approaches.

So if we look back we can see a lot of confusion. The past is as variable as the future, although for different reasons—and we tend to construct or project it to suit our present needs. It is at least interesting that, as information has accumulated about the mentation of other animals, there has been no significant reduction in the ways in which that information is interpreted— in the range of opinions about what goes on in our heads and in those of our mammalian and avian kin. It seems unlikely that this diversity can be explained in terms of reductive rightness or wrongness, any more than can analogous diversity in the past. So maybe it would produce additional clarity, as well as additional confusion, to try to recognize the religious, philosophical, political, and sociological commitments that inevitably supplement the scientific ones in current debate.

Notes

This essay is based on a paper originally presented at the 1999 meeting of the Society for Integrative and Comparative Biology, as part of a symposium on animal consciousness.

1. Tyson, *Orang-outang*, n.p.
2. *Historical Miscellany*, 3:288–89.
3. Peel, *Zoological Gardens of Europe*, 205–6; *In Memory of Consul*, n.p.
4. Romanes, *Animal Intelligence*, 439.
5. Romanes, *Mental Evolution in Animals*, 352, inset.
6. J. O. French, "Inquiry Respecting the True Nature of Instinct," 2, 9.
7. Edward P. Thompson, *Passions of Animals*.
8. Pennant, *History of Quadrupeds*, 1:iv; Wood, *Zoography*, xvii.
9. Quoted in Hull, *Darwin and His Critics*, 164–65.
10. Darwin, *Descent of Man*, 919.
11. Mayr, *Growth of Biological Thought*, 97.
12. Boakes, *From Darwin to Behaviourism*, chap. 4.
13. Romanes, *Animal Intelligence*, vii; idem, *Mental Evolution*, 234–35; idem, *Life and Letters*, 15.
14. Morgan, *Animal Behavior*, 141–43; idem, "Limits of Animal Intelligence," 227–28.

— 5 —

Plus Ça Change

Antivivisection Then and Now

The word *vivisection* has an old-fashioned ring, and *antivivisectionist* is even more suggestive of quixotic Victorian crusades. Yet speakers at a 1983 conference entitled "Standards for Research with Animals: Current Issues and Proposed Legislation" invoked both terms frequently.[1] As its title suggests, the focus of the conference was hardly antiquarian, and the unusually large audience it attracted testified to the timeliness of the topic. Several hundred researchers and administrators from major universities and hospitals, pharmaceutical companies, and government agencies, as well as a sprinkling of humane society members, came to hear two days of panels and discussions. There was a flurry of last-minute registrations, and the organizers had to schedule extra sessions of the most popular workshops.

The sponsor of the gathering, PRIM&R (Public Responsibility in Medicine and Research), a nonprofit organization based in Boston, regularly mounts conferences about ethical issues in biomedical research. Founded in 1974 by a group of scientists and lawyers, PRIM&R considers itself an advocacy group for "appropriate and ethical research that will both improve the quality of life and benefit society" at a time when "public sentiment towards research has grown increasingly hostile."[2] The topic of the October conference was chosen for its timeliness in this context of concern, but even so, the organizers were surprised by the intensity of interest in the fate of experimental animals.

The prospectus announced that the conference would "explore the pres-

---

"Plus Ça Change: Anti-Vivisection Then and Now" originally appeared in *Science, Technology, and Human Values* 9, no. 2 (Spring 1984): 57–66.

ent conduct of research with animals in this country with emphasis on the ethical, scientific, medical and administrative aspects of such research." The actual agenda discussed by most of the speakers, however, was considerably more focused. They were primarily concerned (as was their audience, on the evidence of question periods and hallway conversations) with the threat of increased lay interference in the design and execution of scientific research. Although the tone of their discourse was overwhelmingly temperate and respectful, most of the scientific speakers assumed that humane criticism was rooted in either ignorance (lack of scientific understanding) or false priorities (sentimental preference for animals over humans). They also assumed that the movement to regulate the experimental use of animals was new, even faddish, a product of the expanded liberal sensibilities of the 1960s and 1970s. Several speakers claimed that the publication of Peter Singer's *Animal Liberation: A New Ethics for Our Treatment of Animals* in 1975 had provoked a sudden rise in consciousness and activism.[3]

Both assumptions tend to underestimate the seriousness of the animal protection movement, the strength of the response it evokes among the lay public, and the depth of its historical roots. That is not to suggest that current resistance to scientific experimentation on animals is a linear descendent of the nineteenth-century agitation that prompted the first British legislation about the scientific use of animals.[4] Yet it is revealing that several contemporary organizations preserve the term "anti-vivisection" in their names, despite its overtones of crankiness and lost causes.[5] And the terms of the current debate echo those used a century ago with sometimes startling precision, although the science involved has changed beyond recognition and most animal advocates now base their convictions in philosophy rather than in religion. The most self-conscious animal advocates recognize this historical analogy.[6] If scientists shared this awareness, they might gain a more accurate and more worrisome understanding of the confrontation in which they find themselves engaged.

The beginnings of the animal protection movement in England (which led America in this respect well into the nineteenth century) can be traced to the end of the eighteenth century. Scattered testimonials to earlier humane concern for the welfare of beasts exist, but until the Romantic period, when English sympathies widened to include primitive peoples, the poor, women, and other previously disregarded groups, such concern was apt to be regarded as eccentric, even self-indulgent. As late as the 1780s, it was rumored that Humphrey Morice, a member of the Privy Council, had provided for his thirty aged horses and dogs not in the main body of his will, but in a secret codicil included in a letter to a friend, presumably because he feared posthumous ridicule.[7] Morice was probably unnecessarily apprehensive, however; by 1780, thinkers as diverse as Evangelical clergymen and Jeremy Bentham were advocating the right of animals to humane treatment.[8]

The first attempts to create a legal basis for enforcing this new sensibility quickly followed. A bill to abolish bullbaiting was narrowly defeated in Parliament in 1800, and a broader bill to prevent malicious or wanton cruelty met a similar fate in 1809. Finally, in 1822, a bill "to prevent cruel and improper treatment of Cattle" became law. Its provisions were comparatively narrow and feeble, and applied only to horses, sheep, asses, cows, and steers. Bulls were excluded, along with dogs, cats, pigs, goats, birds, and wild animals.[9] Nevertheless, the act established the authority of government to regulate the treatment of privately owned animals.

Whether the government would exercise its new authority was another question, especially since there was not yet any regular constabulary. To ensure that the act was enforced, its sponsor, Richard Martin, with a diverse group of supporters including Evangelicals and utilitarians, politicians, and clergymen, founded the Society for the Prevention of Cruelty to Animals (SPCA) in 1824. From the beginning, the society employed a corps of inspectors, whose job was to apprehend and prosecute lawbreakers.[10] It soon became the leader of respectable humane opinion. As early as 1835 it received the royal patronage of the Duchess of Kent and the then Princess Victoria, and in 1840 Queen Victoria allowed the society to add the prefix *Royal* to its name.[11]

The humane movement and the antivivisection movement were not identical (as they are not now), although their concerns overlapped. The RSPCA, for example, was more inclined to admit the claims of science than were organizations exclusively concerned with animal experimentation. Even its prospectus of 1824, before the issue had become highly charged, equivocated: "However justifiable it may be to conduct certain experiments of a painful nature, under the control of a benevolent mind, with the view to determine some important question in science, not otherwise attainable, yet all must agree that Providence cannot intend that the secrets of Nature should be discovered by means of cruelty."[12] Half a century later, when public concern about experimentation on live animals was at its height and Parliament was preparing to consider regulatory legislation, John Colam, then secretary of the RSPCA, offered similarly ambivalent testimony to the Royal Commission on the Practice of Subjecting Live Animals to Experiments for Scientific Purposes. He recommended that no painful experiments on live animals be permitted, but he recognized the value of scientific research and praised the humanitarian concern of the majority of British experimenters.[13]

Although many individual RSPCA members were enthusiastic antivivisectionists, many other members of this large, wealthy, and pragmatic organization were eager to support experimental scientists in research that would improve the lives of animals as well as people. The Cruelty to Animals Act of 1876—which mandated government surveillance of animal experiments through annually renewed licenses granted to experimenters, but did not specifically prohibit any research activities—was a disappointment to the RSPCA. Nevertheless, the society soon returned to its primary concern with everyday cruelty; by 1881, a survey conducted by the *Zoophilist,* an antivivisection periodical, found that almost no local RSPCA chapters were actively antivivisectionist.[14]

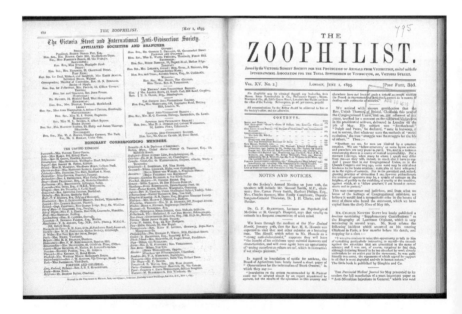

The Zoophilist, 1 June 1895.

The Cruelty to Animals Act of 1876 assuaged the fears of genuine antivivisectionists hardly at all. They tended to be more idealistic than humane society members, more principled and systematic in their objections to animal experimentation, more radical in their criticism of the scientific enterprise. Henry Salt, the philosopher of the late Victorian antivivisection movement, viewed "the awful tortures of vivisection" as the logical outcome of both the lack of reverence for nature required by the analytic methods of science and the arrogance of scientists who, "however kindly and considerate in other respects, have never scrupled to subordinate the strongest promptings of humaneness to the least of the supposed interests of science." Under these conditions, it was as "inevitable that the physiologist should vivisect as that the country gentleman should shoot."[15]

Salt was unusually restrained in expressing his feelings. The rhetoric of George Richard Jesse, an eccentric dog lover whose tactics did not attract much support from fellow antivivisectionists, was more representative. He founded and headed his own pressure group, which he dubbed the Society for the Abolition of Vivisection or Putting Animals to Death by Torture under any Pretext Whatever (later renamed the Society for the Total Abolition and Utter Suppression of Vivisection); and he referred to governmental and scientific attempts to combat rabies as "the brutal massacre of innumerable innocent, intelligent and affectionate animals."[16] For many antivivisectionists, humanitarian revulsion from pain was intensified by religious scruples. They could not believe that a benevolent God would make pain the means of good. So they shared the anguish expressed by Anna Kingsford, a leading figure in the movement as well as both a doctor and a mystic: "For what becomes of the belief in a good and all-compassionate God, if men are to be taught that the way to knowledge and healing involves deeds which hitherto have been supposed to characterize only the worst and wickedest of cowards?"[17]

If the Cruelty to Animals Act of 1876 represented a legislative victory for scientists, it did not guarantee that the public would sympathize with their point of view. For several decades afterwards, vigorous and emotional antivivisection campaigns continued to arouse widespread sympathy. The English often seemed to prefer the claims of animals who were victims of experimental cruelty to those of humans who might be helped by the resulting scientific advances. This tendency was particularly marked when the interests of dogs were at stake. For this reason, the Victorian debate over rabies control stirred especially strong public support for the antivivisectionist position, even though hysterical fear of the disease itself was also common. Throughout the debate, scientific progress was only one of several competing values, and not the one that commanded the most heartfelt popular allegiance.

Although rabies had been known in England for centuries, only scattered individual cases had been reported until about 1735, when the incidence of rabies seemed to increase noticeably, especially among dogs. Outbreaks occurred every decade or so, and because these were concentrated in London and in the industrial parts of Lancashire, increasing numbers of people and animals were at risk.[18] In 1877, the worst year on record, seventy-nine people died in England and Wales. Although such casualties were not particularly high in comparison with those resulting from other hazards—common killers like tuberculosis, typhoid, and diphtheria took a much higher toll, and according to one informal calculation even murder was a fate ten times more likely to befall the Victorian Englishperson—they sparked public concern.[19] Throughout the nineteenth century, Parliament regularly considered bills to prevent the spread of rabies and to compensate people bitten by rabid dogs. The disease was investigated by government committees in 1830, 1887 (twice), and 1897.[20] Despite the almost total absence of reliable knowledge about rabies (or hydrophobia, as it was popularly known), a steady stream of books and articles carried disinformation to doctors, pet owners, veterinarians, and concerned citizens.

Rabies in humans was associated with the bite of a rabid dog, but not every such bite resulted in rabies, nor did every human victim remember being bitten by a dog. This lack of data might have reflected the fact that the incubation time for rabies is usually about six weeks, but can extend to over a year; or it might have reflected the fact that although a bite is the most likely means of introducing the rabies virus into the human bloodstream, it is not the only way. There was even more confusion about the genesis of the disease in dogs. During most of the nineteenth century, the prevailing opinion was that although any animal (including man) might become rabid after being bitten by a rabid animal (generally a dog), rabies could also be "spontaneously generated" in dogs that were exposed to a variety of unpleasant circumstances. Depending on which authority was consulted, these included overfeeding, sexual frustration, hot weather, cold weather, wet weather, thirst, hunger, confinement, terror, and pain.[21]

LEX TALIO-NIS.
(BY A SAD DOG.)

WHY clap dogs into muzzles
In this time of parching drought?
'Tis the knottiest of puzzles,—
From which we can't get out.

It is not yet the dog-days,
And even if it were,
We 're not more giv'n to rabies, then,
Than in winter time, we swear.

Punch cartoon, 1868.

Because the theory of spontaneous generation meant that rabies could arise in any dog at any time, it undermined the plausibility of one effective official response to a rabies outbreak, which was to require that all dogs within the affected area be confined, leashed, or muzzled. Nevertheless, authorities continued to impose such measures, in part because nothing else was available and in part because experience showed that they worked. In fact, the more stringently they were enforced (meaning that the more noncompliant dogs were destroyed), the more effective they were.[22] Except at the height of a local epizootic, however, dog owners resisted these strictures. They were troublesome to comply with, they interfered with the dogs' freedom and pleasure, and the muzzles were painful. It was suggested by a physician who did not believe in microbes that the use of the muzzle actually "developed the rabic matter in the blood of dogs."[23]

When it came, the theoretical explanation of why muzzling and confinement worked was no more persuasive to many citizens than the practical demonstration of their efficacy. After experimenting on dogs for several years, Louis Pasteur inoculated the first human being with rabies vaccine in July 1885. News of this and subsequent successes aroused interest in London, and the next year the Local Government Board appointed a committee to visit his laboratory and scrutinize his results. One member of the committee was George Fleming, the most distinguished British veterinarian of his period and the author of *Rabies and Hydrophobia*. In that authoritative work, published more than a decade earlier, Fleming had rather reluctantly admitted the possibility of spontaneous generation, because he had no scientific counterexplanation of the many reported cases in the medical and veterinary literature. After reviewing Pasteur's work, however, he concluded that "the malady never arises spontaneously."[24] The fact that a vaccine made from the spinal cords of infected animals produced, in most cases, resistance to subsequent, more serious exposure meant that all cases of rabies were transmitted by bite, scratch, or other contact with the saliva of a rabid animal.

The public health implications were clear. Rabies could be controlled in either of two ways. Mass immunization of potentially vulnerable animals was considered impracticable by Fleming and his fellow committee members, because they did not think it would be possible to overcome the reluctance of dog owners.[25] The other alternative was suggested by Pasteur himself. Because the sea presented an insurmountable barrier to migrating wild animals (or roving domestic ones) that might carry rabies, he had been puzzled by British interest in large-scale vaccination. "You do not require it in England at all," he told Victor Horsley, a surgeon and the secretary of the Local Government Board committee. "I have proved that this is an infectious disease: all you have to do is to establish a brief quarantine covering the incubation period, muzzle all your dogs at the present moment, and in a few years you will be free."[26]

A large part of the British public greeted the news that it was about to be released from a terrifying scourge not with gratitude and relief, however, but with outrage. They viewed the carefully weighed and formulated conclusions of the most distinguished scientists and the highest government officials as evidence of a renewed conspiracy against animals (especially dogs) and their advocates. The very prestige of the endorsement seemed an index of the seriousness of the threat. Antivivisectionists led a coalition that ranged from sentimental dog lovers to antimodernizing members of the medical profession in a sustained attack on Pasteur, who became a symbol of all that was most objectionable in experimental science.

To discover the infectious nature of rabies, Pasteur had injected numbers of healthy animals with matter from infected animals, and observed the formerly healthy animals die horrible deaths. To produce the vaccine, he used this same technique to maintain a supply of rabid rabbits. Thus, Pasteur's main laboratory, as well as the Pasteur Institutes that were quickly established across Europe to dispense rabies treatment, were seen to inflict incredible suffering on animals, even if they offered relief to people.

Antivivisectionists also disputed the scientific consensus that Pasteur's theory was correct and that his vaccine worked. Some criticized from a parascientific point of view, expressing concern about the secrecy surrounding his experiments (a concern that had also prompted the Local Government Board committee to duplicate them before endorsing Pasteur's results) or characterizing his work as "very brilliant, but . . . not solid." Or their denial could take the form of name-calling: "M. Pasteur's excuses and inconsistencies are of exactly the same kind as those of the vulgar quack to be found in every country market-place and every country fair."[27]

Some critics claimed that the institutes spread disease rather than cured it. Then as now, a small number of the patients who underwent the series of inoculations after being bitten by rabid animals subsequently died of rabies. Such cases were eagerly exploited as evidence that Pasteur and his scientific supporters were no better than murderers. The institutes were also said to threaten the health of those who lived nearby. The stables of infected animals maintained by Pasteur's laboratories—"cesspools of disease" in the words of one antivivisectionist—were independent hazards. "Wherever a Pasteur Institute has sprung up," charged Thomas Dolan, a Yorkshire physician, in 1890, "the number bitten by rabid dogs has increased."[28]

The scientific case against Pasteur was never strong, and those who tried to make it were driven quickly to forlorn claims that rabies might arise spontaneously in people as well as in animals, that Pasteur cured only patients whose bites had already been cauterized (the only even moderately effective treatment known before the vaccine) or who had not really been infected, or that perfectly effective folk cures existed in remote rural districts.[29] The weakness of this evidence did not diminish the fervor or the convictions of antivivisectionists, however, because these were fueled from another source.

The real objection to Pasteur was more radical and profound, based on moral revulsion rather than flaws in experimental method. Antivivisectionists were ready to condemn the whole physiological and pharmacological enterprise of which Pasteur was only the most visible and obnoxious representative. When they asserted "on moral grounds" that "a system founded . . . on horrible cruelty, could not in a moral universe be fraught with any good," they implicitly rejected the universe of experimental science, in which goodness was not a measure of truth. When they claimed that "the natural heart, as well as the educated instincts, of Englishmen, at least, rise up in arms against the official formulation and recognition of such a system," they challenged the legitimacy of scientifically based public health policy.[30] The breadth of these challenges attracted many who were not especially interested in the technical aspects of medical research but instead were eager to commit themselves publicly to the priority of higher, nonmaterialistic principles in spheres dominated by pragmatism.[31]

Pasteur's theory also indirectly provoked a less exotic and probably more widespread resistance. Although they were less ideologically self conscious than the antivivisectionists, the dog owners who resented muzzling and quarantine similarly preferred emotional satisfaction to improved public health. The most respectable animal-oriented organizations, such as the Kennel Club and the RSPCA, endorsed official antirabies measures, but other groups, like the Dog Owners Protection Association, characterized such measures as "vexatious, tyrannical, and arbitrary interference on the part of the authorities."[32] There was enough grassroots opposition to inhibit local enforcement. Municipal and county authorities, responding to the wishes of their constituents, were reluctant to declare rabies outbreaks and activate control mechanisms. For this reason, when the Board of Agriculture was created in 1889, it was given the power to impose antirabies measures in any area it deemed threatened. The results were dramatic. In 1889, there were 312 reported cases of rabies in dogs and 30 registered human deaths; after four years of central enforcement, the numbers had fallen to 38 and 6. Assuming that this demonstration would persuade even the most recalcitrant local authorities, the government returned responsibility for rabies enforcement to them in 1892. The number of cases soared immediately and the Board of Agriculture had to take over again in 1897.[33] In 1902, Britain was declared free of rabies, but even this complete practical vindication did not justify the board's policy in the view of many dissidents, who continued to protest until the bitter end.

It is hard to interpret the suppression of rabies in Great Britain as a conventional scientific success story. Parallel to the discovery of its treatment and cause, and to the implementation of effective public health measures based on that discovery, ran a consistent rejection of not only the research and the policy but also the moral assumptions on which they were based. Antivivisectionists valued purity more than truth, at least as it was defined by late-nineteenth-century secular authorities; they preferred preventing sinful aggression (the torture of animals and, more profoundly, the violent scientific prying into God's creation) to saving lives.

The vigorous activity of antivivisection groups into the 1890s demonstrated the wide appeal of this subversive reordering of priorities. But the antivivisection movement collapsed suddenly in the first part of the twentieth century. The example of rabies suggests the reason. Although many people were willing to equivocate as long as they possibly could, only the most adamant opponents of experimental science would have been willing to deny themselves achieved freedom from a terrifying disease for the sake of principle. The discovery of the diphtheria antitoxin in 1894, which promised to save thousands of lives each year and which would not have been possible without experiments on live animals, was a decisive blow.[34] Antivivisectionism lost its ability to mobilize public sympathy and came to occupy a position on the outer fringes of respectable opinion.

Nevertheless, public sympathy did not disappear; it remained latent, ready to be reactivated. The late-nineteenth-century antivivisection movement had not exactly been defeated on its own terms; no one had proved that the scientific view of the world was morally preferable. Instead, it had succumbed to the overwhelming pragmatic achievement of the opposition. Continuing advances in immunology and other areas protected biomedical research from antivivisectionist protest for much of the first part of the twentieth century, as did the prestige enveloping the entire scientific enterprise. Lately, however, the benefits of research have become less obvious while the dangers have become more evident, and suspicion has mixed with admiration in the public attitudes toward science. Once again, many citizens have begun to judge science according to their own moral standards, rather than accepting the measures of professional achievement that scientists apply to themselves. And experimentation on animals has again become a touchstone for these opposing points of view.

After a century of scientific progress and social change, the alternative positions have changed remarkably little. In 1875, Charles Darwin reluctantly joined Thomas Huxley's scientific lobby against the legislative efforts of antivivisectionists because, although he was no vivisector himself and he abhorred inflicting suffering, he believed that the potential contributions of the flowering science of physiology outweighed his personal repugnance.[35] Current exponents of this position often stress that they are making the same relative judgment.

Sometimes they emphasize the scientific benefits that have resulted from animal experimentation. For example, Dean Franklin M. Loew of the Tufts School of Veterinary Medicine (in his keynote speech to the October 1983 PRIM&R conference) showed a series of slides illustrating the role that experimentation on animals had played in the discoveries of great scientists like Galen, Galvani, Bernard, and Koch; contrapuntally, he showed a cartoon of white-coated scientists in a cage. The cartoon represented what he believed the public wanted to do with animal experimenters, thus eliminating the possibility of future advances equivalent to those of the past. (In 1979, W. D. M. Paton made the same point more sensationally in an article supporting the use of animals in biomedical research. His illustrations were not just graphs and charts, but photographs of people suffering from advanced cases of loathsome diseases that had been overcome as a result of vivisection-based research.[36])

Some scientists at the PRIM&R conference emphasized the other side of their decision—their reluctance to inflict pain on experimental animals. Joseph Spinelli, the veterinarian who headed the animal care facility at the University of California's medical school at San Francisco, described his institution's stringent enforcement of pain standards stricter than those mandated by the federal government and expressed his personal view that no animal should be made to suffer. Like several other speakers, he pointed out that most scientists, whether for emotional or financial reasons, would prefer not to experiment on animals if they could achieve the same results in another way. He recommended a careful evaluation of the cost-benefit ratio of any proposed experiment that would use animals, especially one that would inflict pain.

In his keynote speech, Dean Loew explained that the purpose of the conference was to help the scientific community understand the complexities of the current controversy over animal experimentation and to cope with legislative changes likely to occur in the near future. Such changes are likely because of rising public sentiment for animal protection, fanned and organized by several different groups. As in the nineteenth century, only a few of these groups can be called antivivisectionist. Others, including then and now the various SPCAs, are primarily concerned with the humane regulation of laboratory experiments, especially on primates and domestic animals like dogs and cats.

PRIM&R asked several representatives of groups in the latter category to address its conference, including William H. Curran, director of the Law Enforcement Division of the Massachusetts SPCA, Henry Spira, coordinator of the Coalition to Abolish LD-50 and the Draize Test, and Connie Kagan, chair of the Animal Political Action Committee. Kagan advocated legislation to increase federal control of animal-related research, and Spira has targeted the way that cosmetics manufacturers test the irritating qualities of new products on the eyes of rabbits. The structure of the program implicitly defined both as extremists by putting them on a panel that also included the most truculent representative of what might be called the right-to-research point of view. This was Frankie Trull, executive director of the Association for Biomedical Research.[37] Trull was the only speaker to refer to organized animal welfare groups as "the enemy"; she urged scientists to organize to protect their intellectual freedom from being compromised by regulatory interference and new administrative burdens.

The real extreme of the animal welfare movement—the only part of it that can appropriately be called antivivisectionist—was, however, not represented at the conference. According to a 1982 internal report prepared at Harvard University, the "most diligent, tactical and clear thinking" groups within the movement are not regulationists, such as those who spoke at the PRIM&R conference, but abolitionists. Although such groups as the Society for Animal Rights and Attorneys for Animal Rights constitute a minority of animal advocates, they have, according to the Harvard report, been disproportionately successful in arousing popular feeling. "Dogmatic" and "professional" in their approach, these organizations value the rights of animal subjects more highly than they do scientific progress. Abjuring "speciesism," they regard the claims of animals as similar to those of human beings, whose rights as experimental subjects have been elaborately protected by law.[38]

Some of PRIM&R's speakers did refer to people who objected strongly to animal experimentation without acknowledging the overriding claims of its scientific necessity and its ultimate benefit to mankind, but such people were not characterized as politically or philosophically sophisticated. Instead, they were presented as untrained sentimentalists who happened to glimpse the operations of a laboratory in the course of their duties—secretaries, nurses, cleaning personnel—or members of the surrounding community who heard garbled accounts of experiments from such impromptu observers. Most speakers supposed that similarly sentimental naïveté was what animated those citizens who wrote anguished letters to their members of Congress. In both the plenary sessions and the smaller conference workshops, many scientific administrators stressed the importance of public education, both within the immediate neighborhood of research institutions and on a national level.

This optimistic interpretation of the resistance to experimentation on live animals may explain the tone of the PRIM&R conference, which was strikingly temperate. Dean Loew announced at the outset that the purpose was not to debate pros and cons, and most speakers seemed to share the sense, expressed by Andrew Rowan, also of the Tufts School of Veterinary Medicine, that everyone was really on the same side. The scientists would prefer not to use animals and the representatives of humane societies actually understood the need for biomedical research. All it will take is a little enlightened compromise to protect animals from abuse without stopping scientific progress.[39]

Rowan dismissed people whose objections could not be disposed of in this pragmatic way as Luddites. Arthur Caplan, a philosopher from the Hastings Institute who focused on the ethical issues, further reassured the conference that thoughtful advocates of both sides—sensitive scientists and sensible antivivisectionists—were not too far apart on some basic questions. They agreed on the need for research, on whether animals are morally equal to people, and on the moral priority of animal ethics issues. The predominant rhetoric of harmony and cooperation gave little explicit indication that the biomedical community was facing what the Harvard University internal report had referred to as a "formidable challenge."[40]

Nevertheless, conference participants seemed to share a sense of being on the frontlines, under attack from a softhearted public easily moved to outrage. Unless it could be reeducated, the public would continue to constitute a potentially serious impediment to research, especially when public opinion was reflected in legislative action. Both academic researchers and private-sector scientists worried about the additional regulation that might result from heightened legislative concern over animal experimentation, and some speakers suggested that current procedures were oppressive.

Conference participants frequently accused the press of stirring up public sympathies by sensationalistic reporting. The most acute resentment of media coverage surfaced in the workshop entitled "Public/Media Perception of Animal Research Issues," which attracted approximately forty participants on the last afternoon of the conference. Because most animal research laboratories are vulnerable to antivivisectionists who want easy targets for adverse media publicity, scientists and research administrators in this session acknowledged the sense of threat and antagonism that had been submerged in most of the plenary presentations. Several participants advised that scientists who ventured to speak directly with the press, rather than using public relations experts as intermediaries, were asking for trouble.

A quick survey of recent periodical indexes suggests that scientists exaggerate the extent to which the press subjects their research to constant and hostile scrutiny. Journalistic attention to animal experimentation had, in fact, been rather intermittent in the years preceding the conference. The *Readers' Guide to Periodical Literature* reported less than one entry per month under the heading of animal experimentation in the period from March 1981 to September 1982; items appeared with only slightly greater frequency in the *New York Times Index* for 1982 and the first part of 1983. (On the other hand, the *Boston Globe* for 4 October 1983 headlined the decorous PRIM&R conference as a "clash.")

Nevertheless, throughout the discussion, the press was blamed for the attitudes of the general public. It was assumed that these attitudes would be inimical to the interests of science unless they were manipulated carefully. And, in some cases, it was feared, no matter how legitimate the research, no manipulation would be possible. Undisciplined sentimentality and indifference to or incomprehension of scientific methods and goals made it unlikely, for example, that Americans would ever approve of the shooting of anaesthetized dogs so that scientists could study gunshot wounds; experimenters at the Uniformed Services University of the Health Sciences found this out in July 1983—to their distress and to the embarrassment of the Pentagon. The principal investigator of that experiment, who took part in the workshop, placed the blame for the "media-inspired" furor on an antivivisectionist "member of the board" of the *Washington Post*.

In the debates on animal experimentation, as on other issues, scientists tend to dismiss those they cannot persuade, whether the opponents are soft-hearted pet lovers or philosophers who argue that people have no right to exploit other species for their own benefit. Despite the denials of many speakers at the PRIM&R conference, there are two sides to the animal experimentation issue; when push comes to shove, these viewpoints may be irreconcilable. Although many scientists and members of humane organizations are working for compromises that will allow research to proceed without inflicting suffering on animals, such compromises accept the premises of one side while rejecting those of the other. They recognize scientific progress or freedom of investigation as a good that must, in some circumstances, override the rights or feelings of animal subjects. To the extent that ordinary sober citizens do not endorse this moral calculus, as many did not in the Victorian era, antivivisectionism will flourish. In a period of public recoil from many of the products of science and technology, the prospects for a speedy ebb of antivivisectionist sentiment are uncertain.

Notes

1. The conference was held in Boston, 3–4 October 1983.

2. Quoted from the package of materials given to registrants at PRIM&R's 1983 conference.

3. Singer's book does represent a recently renewed interest in the ethics of man's relationship to animals among academic philosophers. Other examples include the earlier Godlovitch and Godlovitch, *Animals, Men, and Morals;* and Regan, *Case for Animal Rights.*

4. For an excellent analysis of the politics surrounding this legislation, see Richard D. French, *Antivivisection and Medical Science,* chap. 5.

5. For example, the American Anti-Vivisection Society and the New England Anti-Vivisection Society in the United States and the National Anti-Vivisection Society and the Scottish Society for the Prevention of Vivisection in Great Britain.

6. One indication of this historical consciousness is the recent republication by the Society for Animal Rights of Henry S. Salt's *Animals' Rights Considered in Relation to Social Progress,* which originally appeared in 1892. The new edition includes a preface by Peter Singer.

7. Harwood, *Love for Animals,* 216.

8. Stevenson, "Religious Elements," 148–50; Harwood, *Love for Animals,* 168–71; Fairholme and Pain, *Century of Work for Animals,* 10–12.

9. Turner, *Reckoning with the Beast,* 39–40.

10. Fairholme and Pain, *Century of Work for Animals,* 53–59; Harrison, "Animals and

the State," 798–99. A synopsis of the cases prosecuted by these inspectors occupied a prominent position in the *Report and Proceedings of the Annual Meeting of the Society for the Prevention of Cruelty to Animals,* which was printed and distributed to members beginning in 1832.

11. Fairholme and Pain, *Century of Work for Animals,* 71–72, 89.

12. Ibid., 149.

13. Richard D. French, *Antivivisection and Medical Science,* 102–3.

14. Ibid., 84.

15. He also objected to shooting. See Salt, *Animals' Rights Considered,* 92–94.

16. Jesse, "Publications of Vivisection" and "Publications on Vivisection, 1875–1883," the first of a series of eight scrapbooks that Jesse amassed on the topic; *Standard,* 10 November 1886.

17. Kingsford, *Pasteur,* 28.

18. Hole, "Rabies and Quarantine," 244; Kaplan, "World Problem," 7–8; Fleming, *Rabies and Hydrophobia,* 34.

19. United Kingdom, Parliament, *Report from the Select Committee of the House of Lords,* 178; Arthur Shadwell, "Rabies and Muzzling," *National Review* 15, no. 86 (1890): 230–31.

20. The committees included the Select Committee on the Bill to Prevent the Spreading of Canine Madness (1830); the Committee appointed by the Local Government Board to inquire into M. Pasteur's Treatment of Hydrophobia (1887); the Select Committee of the House of Lords on Rabies in Dogs (1887); and the Departmental Committee to inquire into and report upon the working of the Laws relating to Dogs (1897).

21. United Kingdom, Parliament, *Report from the Select Committee of the House of Lords,* 31, 206; *Bazaar, the Exchange and Mart,* 8 December 1877; Kingsford, *Pasteur,* 20; *Times* (London), 10 October 1877.

22. United Kingdom, Parliament, *Report from the Select Committee of the House of Lords,* 67, 93; Royal Society for the Prevention of Cruelty to Animals, *59th Annual Report,* 84.

23. United Kingdom, Parliament, *Report from the Select Committee of the House of Lords,* 111.

24. Fleming, *Rabies and Hydrophobia,* 92–93, 110; United Kingdom, Parliament, *Report from the Select Committee of the House of Lords,* 80.

25. United Kingdom, Parliament, *Report of a Committee appointed by the Local Government Board,* vii.

26. Horsley's recollections appeared in the *British Medical Journal,* which is quoted in Lise Wilkinson, "Development of the Virus Concept," 30. It is interesting that none of these authorities considered the possibility that cats or wild animals would continue to harbor the disease and infect even a muzzled, registered, and quarantined dog population. The 1887 House of Lords committee received evidence about both rabid cats and rabid foxes. Nevertheless, a program focused exclusively on dogs ultimately succeeded

in eradicating rabies from Britain, which is strong prima facie evidence of the absence of sylvatic infection. According to a modern veterinary scientist, however, "because of the wandering habits of rabid dogs and the extreme susceptibility of the fox to experimental infection, it seems very curious that sylvatic infection has not occurred" (Hole, "Rabies and Quarantine," 244).

27. Kingsford, *Pasteur*, 7; Clarke, *Pasteur Craze*, 3.

28. Jesse, "Publications of Vivisection," n.p.; Dolan, *Pasteur and Rabies*, 76.

29. Kingsford, *Pasteur*, 19–20; Clarke, *Pasteur Craze*, 1, 6; Dolan, *Pasteur and Rabies*, 3.

30. Clarke, *M. Pasteur and Hydrophobia*, 8; Kingsford, *Pasteur*, 28.

31. Richard D. French, *Antivivisection and Medical Science*, 237, 406.

32. *Kent Dog-Owners and the New Muzzling Order*, 5.

33. United Kingdom, Parliament, *Report of the Departmental Committee*, 6–7; John K. Walton, "Mad Dogs and Englishmen," 229–30.

34. Turner, *Reckoning with the Beast*, 115.

35. Darwin, *Life and Letters*, 3:199–203; Richard D. French, *Antivivisection and Medical Science*, 70–76.

36. "Animal Experiment and Medical Research."

37. The more than two hundred members of the Association for Biomedical Research include chemical and pharmaceutical companies and animal breeding laboratories, as well as universities, hospitals, and other nonprofit institutions. One of the group's goals is to monitor proposed legislative and regulatory changes that affect the use and supply of research animals.

38. "The Animal Rights Movement in the United States: Its Composition, Funding Sources, Goals, Strategies and Potential Impact on Research," an internal report to the University prepared by Harvard University's Office of Government and Community Affairs, and based on research by Phillip. W. D. Martin. The report was reprinted by the Society for Animal Rights, Clarks Summit, PA, in 1982; the quotations are from pp. 5, 9, and 2 of the reprint.

39. For a more elaborate statement of this view, see Rowan and Rollin, "Animal Research—For and Against."

40. "Animal Rights Movement in the United States," 3.

— 6 —

Mad Cow Mysteries

When, in March 1996, mad cow disease arrived in the British headlines to stay, not everyone was shocked. Faithful listeners to "Farming Today," which airs just after 6:00 every morning on BBC Radio 4, understood that the spread of mad cow disease might lead to a variety of terrible consequences, human illness among them. Anyone seriously concerned with British cattle—or, indeed, anyone who had paid attention to the intermittent but increasingly troublesome reportage of the previous decade—would have become quite apprehensive about the subject. As a cultural historian of Britain, who has a particular interest in domesticated animals and who spends a good deal of time in British libraries, I belong to both of these audiences. So I was dismayed but not surprised when the British government dramatically abandoned its previous reassuring line and admitted that whatever was making the cattle stagger and collapse might possibly be transmitted to humans with similarly fatal effects.

Less bovine-oriented consumers of news had apparently been responding less attentively over the years. The immediate response to this reversal of official doctrine indicated that most of Britain's population and many of its trading partners had not been prepared for its most alarming implications. Many nations, including New Zealand, Egypt, South Korea, and South Africa, as well as Britain's fellow members of the European Union, suspended the

---

"Mad Cow Mysteries" originally appeared in the *American Scholar* 67, no. 2 (1998): 113–22 (Copyright © 1998 by Harriet Ritvo), and is reprinted by permission of the publishers.

importation of British beef. At home, consumers expressed their concern with their feet, leaving steaks and joints to languish on the shelves of supermarkets and butcher shops. McDonald's restaurants in the United Kingdom quickly announced that their burgers would henceforward be innocent of native beef.[1]

The decisiveness of this reaction (as well as the complaisance that preceded it) could be interpreted as testimony to the influence and credibility of the British government. But a more modest and more complicated assessment would probably be nearer the mark. After all, the official announcement hardly signaled a novel source of concern. Even during the decade of public denial, mad cow disease (technically known as bovine spongiform encephalopathy, or BSE, because it afflicted cattle by causing their brains to become spongy) had lurked on the disturbing fringes of British consciousness. The possibility that it might pose a threat to the health of beef-eating people had been recognized, at least in theory, from the initial observation of its alarming symptoms, even if no one knew what was making the cattle sick, or how they had been exposed to the unknown agent. This baseline anxiety was intermittently intensified by reports that the BSE agent seemed to be able to jump species. In 1990, for example, the disease was identified in kudu, oryx, eland, myala, and gemsbok (all antelopes, which are closely related to cattle), as well as in the domestic cat (a carnivore, and therefore only remotely related to cattle). Further, mad cow disease was recognized as belonging to a group of lethal neurological diseases—spongiform encephalopathies—that included scrapie (a common affliction of sheep), kuru (common only among a small group of New Guineans), and Creutzfeldt-Jakob disease or CJD (extremely rare but widely distributed among human populations). Speculation arose that the BSE agent might also cause a newly identified variant of CJD, which afflicted younger people and took longer to run its fatal course. In 1993 the *Lancet* suggested that the death of a farmer from CJD had been caused by an occupational hazard: contact with his BSE-infected herd.

Responses to these inconclusive but troublesome suggestions varied. Australia and the United States banned the importation of British beef and cattle in 1989. Several European countries followed suit in 1990, but quickly reversed their decisions after reassurances from the British government. On the whole, however, such reassurances were not required. Even at home, where BSE-related items appeared most frequently in the news, and where consumption of British cattle products, and therefore exposure to anything harmful that they might contain, was highest, the disease was routinely trivialized. The flurries of alarm that greeted each new revelation tended to die down quickly, although they sometimes left added precautions in their wake, and they inevitably contributed to a latent sense of unease.

Tory politicians (the Conservative Party was in power throughout this period) tended to view mad cow disease through the eyes of the beef industry, as a potential economic catastrophe, rather than as a threat to public health. They instituted regulations reluctantly and implemented them without conviction. They repeatedly downplayed the possibility of transmission to humans, most flamboyantly in 1990 when Minister of Agriculture John Gummer confidently fed a beefburger (because they included beef products of varied derivation, fast-food burgers attracted special suspicion) to his four-year-old daughter for the edification of the media.

His lighthearted attitude was widely shared, by both journalists and the larger public. The first cat to die of BSE was headlined as "Mad Moggy." One of the many websites devoted to mad cow disease featured only a pasture full of loopy cartoon cattle, with revolving pinwheels for eyes. And, although only a few days after the official bombshell of March 1996 the beef coolers at my local Sainsbury's were completely empty, the first explanation that occurred to me turned out to be completely wrong. I guessed that the unpurchased meat had spoiled and been returned whence it had come. But instead, in response to shoppers' initial shocked avoidance, Sainsbury's (along with the other large supermarket chains) had simply cut beef prices in half. Many customers eagerly stocked their freezers with meat that had seemed too dangerous at regular rates.

Perhaps encouraged by official measures and explanations (including the removal of cattle more than thirty months old from the food chain, the exclusion of meat and bone meal from feed destined for farm animals, and the assurance that the heads of slaughtered cattle would join their spinal cords and other organs suspected of concentrating the BSE agent as "specified bovine offal," which could not be unrecognizably reprocessed into other food products), British beef consumption soon returned nearly to prescare levels. But the sense of renewed security enjoyed by some British beef-eaters (certainly not all—among nonvegetarians of my acquaintance, responses also included total avoidance of beef and avoidance only of ground beef, as well as skepticism and fatalistic indifference) ended at the beaches. The European Union imposed a ban on the export of British beef and beef products, not only to other member nations, but throughout the world. As I write in the autumn of 1997, this ban is still in place, although under the newly elected Labor government, as under the old Conservatives, MAFF (the British Ministry of Agriculture, Fisheries and Food) has made repeated and energetic attempts to have it rescinded. Despite the defensive efforts of their governments, Continental consumers have proved less easily mollified than their insular counterparts. Like British consumers, they conceived an instant aversion to beef after the announcement of March 1996, but unlike them, despite the embargo on British beef, the very small number of cases of mad cow disease that had been reported in Continental herds, and the fact that local beef was prominently labeled, with words and even with flags, they did not change their minds. They have resisted both official blandishments and those of their beef industries, and beef consumption has remained far below pre-1996 levels.

It is at least interesting that governments and citizens on either side of the English Channel embarked on such different courses of action. Of course, official responses were heavily influenced by economic self-interest: the British desire to preserve the international markets for its beef and beef products, and the somewhat incompatible desire of other European Union members to avoid having their own beef industries tarred with the brush of British BSE. But cattle farmers and meat processors, and the food industry more generally, were not the only groups with a potential stake in BSE policy, although they were by far the best organized. If mad cow disease was transmissible to other species, including humans, then the general public had an interest in its control and elimination that was more than merely financial. Once the possibility of transmission was seriously acknowledged, governments were self-consciously setting policies with potentially enormous public health consequences, and individuals were making, on their own behalf and that of their children, what might literally prove to be life-and-death decisions. Not even the bottom-line and value-for-money oriented Conservative government conceived its mission so narrowly as to exclude this kind of unquantifiable moral responsibility. It would not have been able (or inclined) to defend a policy that demonstrably sacrificed the health of large numbers of citizens in an attempt to maintain the profits of an industry.

But it is not necessary to posit any such sinister calculation to account for British policy. One reason that so many inconsistent courses of action seemed possible in March 1996 was that decision making was relatively unconstrained by information. The answer to the most important question—whether mad cow disease posed a significant threat to human health—was, and remains, unclear, although the indications have become increasingly troubling, as the number of human cases slowly accumulates and the relationship between BSE and the new variant of CJD demonstrably strengthens. Even the basic facts of the BSE epidemic were and are subject to wide interpretation. For example, between its first official recognition in 1986 and the spring of 1997, 168,382 cases of mad cow disease were identified in the United Kingdom, a figure of impressive but possibly misleading precision. The equivalent figure for non-British BSE underlines the problems implicit in such quantifications. Approximately 400 cases have been reported in other European nations, mostly in Switzerland. This number is surprisingly small, when it is considered that, before 1990, thousands of British cattle, along with large quantities of beef, bone meal, and other cattle products, were regularly imported by those nations. It has been plausibly suggested, especially but not exclusively by politicians protesting the ban on British beef, that the European incidence of the disease has been seriously underreported, in part as a consequence (the reverse of what was intended, but predictable enough, given human nature) of draconian sanctions, such as the destruction of any herd in which a single case appeared, combined with inadequate compensation to the unfortunate farmer who owned it.

Within the United Kingdom, the pattern of infection was equally puzzling, if better documented. Mad cow disease was more prevalent in dairy herds than among beef cattle, more common in England than in Scotland. Some herds, even in areas where it was prevalent, remained free of it. Epidemiological analysis of the first rash of cases identified commercial feed that incorporated meat and other material from sheep and possibly also cattle as the only factor that linked all the outbreaks. It was theorized that the cattle had ingested the BSE agent as they munched the remains of their ruminant brothers and sisters—possibly transmogrified from scrapie-infected sheep, possibly simply transmitted from an already afflicted cow or bull. (As with the AIDS epidemic, which seemed at first to have come from nowhere, retrospective reflection revealed possible earlier cases, suggesting that the disease might previously have been endemic at a very low level.) The feed hypothesis was plausible, and it has provided the basis of the most effective disease control measures. Indeed, the elimination of meat and meat by-products from livestock feed is largely responsible for the dramatic drop in BSE cases in recent years (a drop that had begun before the scare of March 1996); it is possible that within a few years, due to this measure alone, mad cow disease will have been eliminated from British herds. But if contaminated feed was the cause, why had its effect been lethal in some cases and innocuous in others? Given the protracted incubation period of BSE, longer than the lifetime of many beef cattle, could absence of disease be reliably taken to indicate absence of infection? And what of other associations, less powerful but still suggestive, especially exposure to certain organo-phosphorus compounds used in agriculture?

Another set of questions clustered around the elusive agent. The language of epidemics and transmission in which mad cow disease was normally discussed suggested infection, rather than poisoning, although clearly this was no ordinary contagion. Gradually, a scientific consensus emerged that, in the case of BSE, as with the other spongiform encephalopathies, the agent was a distorted prion, a kind of protein molecule found in the brain. (The apparent strength of this consensus was enhanced when Stanley Prusiner, the controversial theorist of prions, won the Nobel Prize for biology.) But although this focus on prions clarified further directions for research, it did nothing to simplify the task of formulating practical responses. For example, it did not explain how animals became infected with mad cow disease. New cases kept appearing after the suspect feed was banned, and only some of these cases could be attributed to farmers' thriftily ignoring public health regulations in order to use up feed that they had already paid for. Could it be transmitted from animal to animal within a herd? From mother to calf? (In the latter case the ban on beef from animals more than thirty months old was beside the point.) Could the infective agent somehow linger in fields where sick animals had been pastured, as was the case with scrapie?

Indeed, because prions were not well understood, and because they turned out to be practically indestructible (in particular, enormously resistant to heat), their implication in the spread of the disease significantly complicated one facet of policymaking. Although the elimination of BSE in cattle offered the obvious ultimate solution to the problem, in the interim it was necessary to protect members of other species, especially humans, from contact with potentially infective material. Removing all cattle from herds with any BSE cases from the food chain, and removing the brain, vertebral column, and other suspect organs from all butchered cattle constituted only a first step. Assuming that it was successfully accomplished (an assumption not always justified), this triage yielded an immense volume of rejected offal and carcasses. The unsavory material had to be disposed of in a sterile or sanitary way so that the distorted prions did not reenter the food chain, or, worse, were not somehow more widely disseminated. Burying was clearly out of the question, as was disposal in coastal waters, where marine mammals might be exposed and where neighboring countries might object. The only practicable solution seemed to be cremation at unusually high temperature, which was made more difficult by the fact that not all incinerators were properly equipped for such procedures. If infected material were inadequately burnt, prions might survive in the smoke and be scattered by the wind, ultimately settling in soil or in water, ready for reingestion. It took more than a year to develop adequate capacity and procedures for this massive slaughtering and incineration; in the meantime contaminated carcasses and offal accumulated in storage.

The toughness of prions also complicated decisions about which beef products posed a risk of transmitting mad cow disease. Although dairy products were generally pronounced safe, opinions varied about gelatin, a substance present in a wide variety of prepared foods, including some, such as packaged cookies or biscuits, routinely eaten by unwary vegetarians. It was much more heavily processed than milk or cheese, but it included material from suspect cattle parts. In April 1996, a government regulation prohibited the use not only of meat but also of bone meal as fertilizer on agricultural land, although use in private gardens and in greenhouses was still permitted (prudent gardeners were advised to wear gloves and masks, however). And the identification of prions as the BSE agent did nothing to allay concerns about their possible cross-species transmission during the previous decade, perhaps (again in analogy to AIDS) seeding a devastating human epidemic, the scale of which would not become obvious for years or decades. In the nightmare scenario delineated by Richard Rhodes in *Deadly Feasts: Tracking the Secrets of a Terrifying New Plague* (1997), the rogue prions have already become so widely disseminated that stopping their further destructive spread will be nearly impossible.

The official announcement of March 1996 did not tilt the odds—or change the reality, whatever it turns out to be—in one direction or the other. But, by altering public awareness, it produced a need for governmental action, both in Britain and in other countries. And if information is not available (indeed, often, even if it is) policies must be formulated on other grounds. This is not to suggest that ignorance was foregrounded as the basis of official decision making. On the contrary, one of the most striking features of the Conservative government's attempt to defend the lighthearted negligence of its previous treatment of the BSE issue was the repeated invocation of "scientific advice" as the touchstone of its policy. Although the specific content of this advice was seldom alluded to, minister after government spokesman referred to it as if it were definitive and authoritative—the quintessential black box—at a time when most scientists concerned with BSE and related matters were emphasizing the prevalence of uncertainty. (The relevant expert community in Britain was in any case smaller than it might have been, owing to earlier official reluctance to fund research that might produce disturbing results.) And the government was aided in this misrepresentation by the fact that, with some notable exceptions, such as the *New Scientist,* science journalism is underdeveloped in Britain, as is the popular audience for such reportage.

Although government policy was proclaimed to conform to "scientific advice," it was more transparently founded on patriotism, which had the advantage of immediate public appeal. From this perspective, the interests of the nation, its citizens, and its cattle industry were happily indistinguishable from one another and from those of the animals themselves. Editorial cartoons portrayed a blighted countryside, heavily shadowed by smoke from a holocaust of innocent cattle—a sad contrast with the beautiful placid herds to be seen throughout rural Britain, but one that neglected to take into account the similarity between BSE-induced culling and ordinary slaughtering from the cattle's point of view. Those inclined to continue or resume their previous dietary preferences could see eating British beef as a declaration of loyalty, even as an assertion of national courage and common sense against the scaremongering of experts and foreigners. Soon after the announcement of March 1996, I went to a Chinese restaurant with a group of people, some of whom, in this enthusiastic spirit, made a point of ordering beef dishes to be shared, and also of noticing who declined to share them.

Simple jingoism, however, does not sufficiently explain the widespread willingness to ignore a risk that, although it might well not exist at all, might also lead to the most dreadful consequences. Food contamination alarms are not routinely viewed with indifference by the British public. For example, a 1988 uproar about salmonella in eggs was famously countered by the heavily publicized consumption of a runny omelet by the prime minister herself. Further, the British tend not to regulate their everyday behavior according to knee-jerk patriotism (certainly, they are less inclined to do so than are

Americans); and the word of the superannuated Tory government had long ceased to inspire implicit confidence, as was confirmed by its decisive defeat at the polls the following year.

But beef and the cattle that produce it holds a special place in British national mythology.[2] Before John Bull was a canine he was a bovine. The wealthy eighteenth-century landowners who devoted themselves to agriculture took special pride in the creation of improved breeds of livestock, such as the shorthorn cattle that remained internationally preeminent for the succeeding century. In this enterprise, the breeders celebrated themselves. It was no accident that the entries in stud books, herd books, and flock books closely resembled those in Burke's and Debrett's catalogues of the aristocracy and gentry. But they also claimed to represent the entire nation at its best, a claim that many ordinary citizens accepted without demur. Many were eager to pay homage to the magnificent beasts produced by breeders when they were exhibited at national livestock shows, by, as one agricultural journalist put it, "closely inspecting, admiring, and tormenting the bullocks sent by Dukes and Earls."[3] The public apparently agreed with Julius Caesar, who had suggested that cattle constituted the true wealth of the Britons, and with an early president of the Royal Agricultural Society, who discerned "in the people of this country a peculiar disposition and talent for encouraging the finest animal forms."[4]

The sustenance such people derived from such cattle was inevitably profound and complex. The roast beef of Old England has for centuries been valued as much more than an efficient source of nourishment. To it was attributed, at least in part, the superiority of the British soldier to his European and imperial antagonists; more generally it has been viewed by observers both within and outside the British polity as an essential component of the national character. As Charles Dickens reflected more than a century ago in *Household Words*, "beef is a great connecting link and bond of better feeling between the great classes of the commonwealth," inspiring respect second only to "the Habeas Corpus and the Freedom of the Press."[5] Further, British naturalists and agricultural experts, chefs and gourmets have tended to agree about their nation's preeminence with regard to the quality of the meat it produces, as well as the quantity it consumes. For example, a Victorian gastronome celebrated native beef while disparaging foreign livestock as follows: "If the poor, half-fed meats of France, were dressed as our cooks . . . dress our well-fed, excellent meats, they would be absolutely uneatable."[6] As a result a deeply

"Johnny Bull and the Alexandrians," by William Charles, 1814.

ingrained tradition of metonymic identification has developed, affectionately acknowledged when the British call themselves beef-eaters, less so when their nearest Continental neighbors call them *rosbifs*.

In any country, the possible condemnation of the national cattle herd would be viewed as an impending disaster, both economic and epidemiological, even if it were not associated with a potentially serious threat to human health. But the time-honored identification of Britain with its beef has made the BSE crisis especially difficult. And the resonance of this patriotic symbolism has not been exclusively domestic. It seems likely that the intransigence with which Britain's "trading partners" in the European Union have defended their import bans, once imposed, as well as the obsessive avoidance of beef by many European consumers, also reflect historic associations and aversions.

At present, the story of mad cow disease is far from over. None of the projected scenarios, from the most reassuring to the most alarming, has yet been definitively eliminated, either in the laboratory or through epidemiological study. Sooner or later, however, the medical and veterinary cases will be closed, and it will be possible to determine who was prudent, who was alarmist, who was foolhardy. But such a restricted accounting should not constitute the final word on BSE, for several reasons. Although mad cow disease happened to strike British herds, it does not seem to have resulted from idiosyncratically British attitudes or practices. Large-scale livestock farming is similar throughout the world, as are the relations between major industries and their national governments. At first glance the reactions of European governments seemed to compare favorably with that of their British colleagues, as they made the health of their citizens their first priority. But when the potential danger to health came from their domestic beef industries rather than from the foreign menace, their inclination was generally similar to that of the Tories. The same point could be made with regard to the United States, which has stalwartly maintained its early ban on British beef products. Officially, according to the Department of Agriculture, the United States has no BSE—but the DOA also goes to the trouble of denying that downer cow syndrome, which afflicts thousands of American cattle each year, is related to mad cow disease. Not everyone is persuaded by this denial.

The shared inadequacy of these responses points to a larger problem that is also shared: the difficulty of dealing with complex technical crises. Such crises by definition require specialized expertise—expertise that strapped or stingy governments probably do not have on the payroll. Often it exists only within the industries or institutions that have the largest financial stake in the outcome; sometimes, as with mad cow disease, such expertise has not yet been sufficiently developed. In either case, before the public becomes alarmed, policy tends to be determined by such interest groups. Afterwards, at least in democracies, the technical crisis is inevitably compounded by a crisis of public relations. Politicians must be seen to act quickly and firmly, whether or not they have any basis for their actions. Such can-do decision making, however impressive and reassuring in the heat of the moment, is unlikely to produce wise policy. On the contrary, it risks catastrophe. It was sobering to realize, during the spring of 1996, that it was impossible for politicians in power to acknowledge that they lacked sufficient information to make certain decisions about how to protect people from the threat of BSE. It was therefore impossible for them to act with appropriate circumspection—to concede that there are times when discretion is indeed the better part of valor. If the protracted scandal of mad cow disease were to lead to a recognition of this problem, that would constitute a faintly silver lining.

Notes

1. This account of recent events was based on information provided by news media, government agencies, and other interested organizations, both in print and electronically. This essay originally appeared in 1998; since then several book-length overviews of the mad-cow-disease crisis have appeared, including Lledo, *Histoire de la vache folle,* and Schwartz, *How the Cows Turned Mad.*
2. See Ritvo, "Roast Beef of Old England."
3. "Fat Stock and the Smithfield Show," *Illustrated London News* 28 (1853): 491.
4. "On the Natural History of the Domestic Ox, and Its Allied Species," *Quarterly Journal of Agriculture* 2 (1830): 195–96; Pusey, "On the Present State of the Science of Agriculture," 17.
5. Quoted in Simmonds, *Curiosities of Food,* 2–3.
6. *Guide to Service,* 12–13.

— 7 —

Understanding Audiences and Misunderstanding Audiences

Some Publics for Science

In June of 1900 the annual show of the Royal Agricultural Society of England took place at York. Since 1839, the RASE shows, which, like the meetings of the British Association for the Advancement of Science (BAAS), migrated from city to city each year, had served as a focus for agricultural achievement, marketing, and sociability. As highly visible and festive occasions, they also reminded citizens without direct connections to the land of the importance of agriculture in the life of the nation. Both farmers and nonfarmers regularly flocked, or so the show's organizers hoped, to watch the judging of numerous classes of cattle, sheep, horses, pigs, and poultry, to admire the latest in agricultural implements and machinery, and to enjoy the holiday atmosphere. But during the 1880s and, especially, the 1890s, these high hopes were seldom realized.[1] Attendance at the shows and, therefore, gate revenue were in decline. The 1899 show at Maidstone incurred a large deficit, and the officers of the RASE counted on the 1900 show, to be held in a populous area well served by rail and road, to recoup their balance. But at least from a financial point of view, York too proved disappointing. Although the rings set up on the Knavesmire show ground—located conveniently near the town center—could be intermittently described as "crowded with spectators,"[2] overall attendance fell significantly below the rosy expectations.

"Understanding Audiences and Misunderstanding Audiences: Some Publics for Science" originally appeared in *Science Serialized: Representations of the Sciences in Nineteenth-Century Periodicals,* edited by Geoffrey Cantor and Sally Shuttleworth (MIT Press, © MIT 2004), 331–50.

Not even fine weather and the conspicuous presence of royalty produced throngs of the required magnitude. The show was held under the presidency of the Prince of Wales, and both he and his venerable mother took first prizes in the shorthorn classes. The prince won the "old bull" competition with a "fine, massive animal" called Stephanis, while the queen's "Royal Duke, . . . a really grand beast, square typical and handsome" was judged the best two-year-old bull, and awarded the overall championship as well.[3] Possibly one problem was the absence of the pig classes, "always a popular feature," which had been canceled on account of an outbreak of swine fever.[4] Or public attention may have been distracted and public spirits depressed by the war news from South Africa, which crowded the RASE show out of the pages of the *Illustrated London News*. In any case, only 87,511 people paid to see the show (admission was 5s., 2s. 6d., or 1s., most expensive on judging day, least expensive as the show wore on), which was, with the exception of Maidstone, the worst turnout in eighteen years.[5] Rather than canceling the Maidstone deficit, the 1900 York show increased it by £3,500.[6]

The picture was not entirely bleak, however. The shorthorn classes "were more attractive than ever" and the Highland cattle had never displayed "finer heads."[7] And if attendance looked relatively sparse to those responsible for the RASE balance sheets, an audience of nearly ninety thousand, many of whom were not professionally engaged in agriculture, was far from negligible in absolute terms. Further, paying customers constituted only a fraction of the show's ultimate audience. The *Illustrated London News* to the contrary notwithstanding, the publicity function of the annual RASE display was enhanced by extensive reportage in newspapers, popular magazines, and periodicals catering to various specialized audiences.

One particular exhibit attracted the attention of journalists and their readers, as it appealed to viewers on the spot. Many show visitors chose to pay an additional sixpence (a charge regretted by the agricultural correspondent of the *Manchester Guardian* as imparting a "particular showmanlike flavour") to view what was noted in the *Times* as a "very popular attraction"— the "zebra hybrids."[8] The show organizers signaled their sense of the significance and attractiveness of these animals by assigning them a very prominent and accessible location. Housed in their own special building on the central axis of the show, the zebra hybrids were close to the Royal Pavilion, the large refreshment area, and the public conveniences.[9] The offspring of a handsome Burchell's zebra stallion named Matopo and mares representing a variety

"Matopo," from James Cossar Ewart, *Guide to the Zebra Hybrids*, 1900.

of domestic horse and pony breeds, these creatures seemed exotic indeed among the familiar farmyard animals competing for RASE prizes, emissaries from the more glamorous world of sideshows and menageries.

The display was mounted, however, in sober scientific terms reminiscent of a natural history museum. For purposes of comparison, the hybrids were accompanied by their mutual sire Matopo, who was praised as "wonderfully quiet and friendly,"[10] along with several of their nonhybrid half-siblings (animals whose fathers were of the same breed—for example an Arab horse or a Shetland pony—as their mothers, who had previously borne Matopo's offspring). Other equines (horses, donkeys, and zebras) were represented either by their painted or by their photographic images, or by their skins. In addition the display featured live pigeons of several recognized breeds—fantails, owls, archangels, barbs, turbits, and jacobins—along with the offspring resulting from various crosses between them, several rabbits whose recent forebears included both domestic and wild animals, and a white cat with her litter of four kittens. The father (and uncle) of the litter was also white. They were included because only half the kittens resembled their white parents in color; the other two recalled one of their great-grandmothers.[11] The press reflected public enthusiasm in lavishing attention on the hybrids; it echoed the explicit tone of the exhibition by emphasizing its scientific purpose, or at least the fact that it had a scientific purpose. Even the *Guardian* correspondent reluctantly admitted that "the animals are worth seeing," although he could not explain exactly why. Instead he grumbled that "while one has no doubt the display is purely biological, one fails to see its practical or utilitarian value."[12]

If this dour journalist had parted with a further shilling and purchased the *Guide to the Zebra Hybrids,* on sale at the exhibit, he might have been less mystified. The booklet's brief introductory note identified its primary audience as "the members of the Royal Agricultural Society of England," and then proclaimed both its own didactic purpose and that of the eye-catching display it described: "to indicate to all interested in the problems of Heredity that, as our knowledge increases, many prevalent views will require to be either discarded or profoundly altered."[13] That is, it was intended to make the scientific understanding of the mechanisms of heredity available to members of the other specialized community most interested in the subject—animal breeders. Indeed, the *Guide* was noticeably modest in characterizing its intended readers. The previous and subsequent public career of the hybrids, both in person and in print, made it clear that their lesson was meant for a much broader set of audiences.

The author of the *Guide* was the owner and creator of the exhibit, James Cossar Ewart (1851–1933), a Fellow of the Royal Society and the Regius Professor of Natural History at the University of Edinburgh. The animals on show at York had resulted from four years of experimentation at his farm at Penicuik, in the Midlothian countryside south of Edinburgh. From the inception of this elaborate research project Ewart had been as interested in publicizing his results as in achieving them. Thus the display itself also represented a culmination of sorts. Before showing them to the diverse national audience gathered at York, Ewart had exhibited his hybrids to a range of local audiences—agricultural, zoological, and simply curious. For example, in 1897 he responded to what a local newspaper characterized as "desire on the part of the general public" by displaying some of his animals in the Edinburgh Cattle Mart, where they attracted "much attention and favourable comment."[14] At the Highland Agricultural Show in 1899, "no exhibits attracted a greater amount of attention and interest than the large number of zebra hybrids" (among their admirers was the Prince of Wales, who consequently proposed them for the next year's RASE show).[15] Specially organized groups could get a more comprehensive view of the experiments-in-progress by visiting the animals at home. In the summer of 1898 one hundred visiting scientists paid 2s. apiece for a day out that included lectures on hybridity and breeding, a tour of the stud, and afternoon tea, as well as a return rail ticket from Edinburgh.[16] Later in the year fifty agricultural students from the Glasgow and West of Scotland Technical College enjoyed a similar program, but without refreshments.[17]

Ewart also put himself on show in the service of publicizing his research, most notably in a series of three lectures entitled "Zebras and Zebra Hybrids" delivered at the Royal Institution in the spring of 1899. Their content was, of course, further disseminated through reports in newspapers and magazines.[18] Ewart was doubtless gratified by both the amount and the variety of attention he received. He respected the power of the press—he subscribed to a cutting service and kept extensive cutting books—and he seems to have been especially conscious of the ability of journalism to engage the attention of readers outside his own natural audience of scientists and intellectually inclined agriculturists. In his quest for sympathetic coverage, he sent photographs of his animals to journalists, and invited them to visit his farm for private views if they found themselves in the vicinity of Edinburgh. This strategy often produced gratifying results. For example, an account in *Polo Magazine* characterized the author's visit as a "pleasure," the zebra Matopo as "one of the finest specimens of the breed," the pony Mulatto as "the heroine of the piece," and their hybrid colt Romulus as "our little striped friend." The piece was illustrated by a photograph of Ewart's ten-year-old son astride a zebra.[19]

Ewart's most comprehensive presentation of his work with the zebra hybrids was in a series of freestanding publications that appeared between 1897 and 1900, of which a small book entitled *The Penycuik Experiments* was the most substantial. Reviews of these publications disseminated Ewart's ideas to a large and diverse body of readers. Or at least, depending on the zoological self-confidence of the reviewers, they disseminated the fact that these ideas existed. Thus the *Sportsman*'s reviewer of *A Critical Period in the Development of the Horse* modestly claimed that "it is almost idle to attempt to deal with Professor Ewart's work at all, when one is so entirely behind him in regard to scientific knowledge," while nevertheless suggesting that any breeder would be "the better for possessing it."[20] Reviews of *The Penycuik Experiments* emphasized the book's relevance to the interests and predilections of various audiences: "a volume which cannot fail to arrest the attention of stock-breeders"

(*Times*); "all biologists will agree in looking with eagerness for more" (*Natural Science*); "of interest to more than one class of readers" (*Lancet*); "offers the public an intellectual treat" (*Morning Post*); "lovers of animals . . . will devour the contents . . . with avidity" (*Scottish Farmer*).[21] The *Irish Naturalist* explained its inclusion of a review of work done in "a small Scotch town" with "no direct connection with Irish natural history," on the grounds that Ewart's experiments were "of general interest and importance," and, moreover, that "several Irish horses" had participated.[22] The *Quarterly Review* published a combined notice of the book and the Royal Society paper, which provided a comprehensive and sympathetic account of Ewart's work in relation both to scientific theory (the review began with a somewhat contrarian invocation of Darwin) and to practical farmyard problems (the deterioration of the English racehorse, the consequences of inbreeding), concluding that "the book cannot fail to attract both the man of science and the practical breeder."[23]

Thus the coverage of Ewart's work at Penicuik was both broad-based and promiscuous. That is, the significance of Matopo and his offspring was discussed in a wide variety of periodicals, appealing to very different audiences, from technical or scientific specialists to casual general readers. Although the emphasis and interpretation varied according to the journal, the variation was less marked than might have been predicted based on their divergent audiences. Subscribers to the *Manchester Guardian* were exposed to much of the same material that the *Lancet* presented to its medical readers. The similarity of reports from periodical to distinct periodical may have had something to do with the difficulty of repackaging—even of paraphrasing—technical material; it certainly meant that the scientific core of Ewart's work was black-boxed for many readers. But this similarity may also have signaled that the focus of interest lay elsewhere. To some extent the fact that Ewart's research on zebra hybrids appeared nearly universally newsworthy was a reflection of his own character and the shape of his career, a large part of which had been devoted to public service and to bridging the gap between zoology and animal husbandry. Further, as was demonstrated at the RASE show, the enthusiastic reception accorded his experiments owed something to the charisma of his experimental animals, and perhaps also to the fact that their lives seemed so different from those of most creatures who were devoted to science. An account of the treatment of vivisection in the nineteenth-century periodical would emphasize the contrast between specialist and general audience coverage rather than convergence. In this context it may be significant that none of the descriptions of the zebra hybrids highlighted the fact that Ewart fatally exposed several of his engaging young animals to the tsetse fly, and most completely overlooked these sacrifices to science.

Although the topic of heredity was of great theoretical concern to biologists, of great practical concern to stockbreeders, and of a good deal of interest to the public at large in the late nineteenth century, these audiences overlapped less frequently than might be imagined. Ewart was unusual in attempting to address them all. Even within the field of zoology, his range of interests was notably broad. As a young researcher, he had collaborated with George John Romanes in investigating the echinoderm locomotor system. His responsibilities as chair of natural history at the University of Edinburgh beginning in 1882 led him to publish on practical anatomy and on the care of collections; he lectured on both vertebrate and invertebrate zoology. (His student audiences may have been among his least appreciative, however. Sir Maurice Yonge, a distinguished invertebrate zoologist who was a doctoral candidate at Edinburgh in Ewart's later years, recalled that in their youthful high spirits they frequently drowned him out, so that only the persistent movement of his walrus moustache indicated that he was still talking.)[24] He worked extensively on the embryology and development of vertebrates, especially the horse and the sheep. As a result of his interest in these fields, the University of Edinburgh established a lecturer's position in genetics (the first in Britain) in 1911. Toward the end of his career Ewart turned his attention to the deciduous feathers of such birds as mallards and penguins.

Ewart was always ready to apply his zoological expertise to pragmatic issues of public concern. As a founding member (1882) of the Fishery Board for Scotland, for example, he made extensive comparative investigations of fisheries policy in both Europe and North America, and also conducted scientific research on the maintenance and propagation of fish stocks. When his scientific focus shifted to hoofed animals, his extraprofessional activities followed suit. In 1897, as part of a commission to survey the horses and ponies of Ireland, he made a set of recommendations that were considered by breed aficionados to have preserved the Connemara pony. He was one of the first vice presidents of the (Royal) Zoological Society of Scotland (the parent organization of the Edinburgh Zoo), and a cofounder of the Park Sheep Society, dedicated to saving seven threatened ovine breeds, including the multihorned Jacob. His work with the zebra hybrids, theoretically driven and even eccentric as it might have seemed, had similarly practical implications. These striking creatures were produced in the course of an experiment designed to test several widespread assumptions about heredity that formed the basis of stockbreeders' choices of sires and dams. By correcting these assumptions and replacing them with more accurate understandings, Ewart hoped to offer breeders the means to improve their matchmaking, and thus the quality of their flocks and herds.

The primary target of Ewart's experiments at Penicuik was the concept of telegony, or "influence of the previous sire" (also termed "infection" or "saturation" depending on the way its supposed operation was explained), as a result of which, it was widely believed by both farmers and zoologists, the father of a female's first child was able to influence her subsequent offspring by different fathers. This belief encouraged breeders to mate each virgin female animal with the best possible male so that his superior qualities would continue to grace her later foals, lambs, or puppies, no matter who sired them; it reciprocally dictated that an inappropriate initial mate (worst of all, one of the wrong breed or of no breed) would cast a pall over a female's entire reproductive career. The evidence for this belief was anecdotal rather than systematic; voluminous rather than scientifically persuasive. The most famous and best-documented instance of telegony (although the documentation consisted mainly of repeated retellings of the original story) was an animal called "Lord Morton's mare," who had flourished eighty years before Ewart displayed his zebra hybrids. She had borne her first foal to a quagga (a less dramatically striped relative of the zebra), and her subsequent foals to ordinary horses. Not only the initial hybrid but all of her later offspring were reported to exhibit striping or barring, especially on the legs, which was interpreted as evidence of the persistent reproductive influence of the quagga.[25] (It should be noted that other interpretive options were available; for example, breeders would have known that such striping is fairly common in horses, especially duns.) And if evidence for the occurrence of telegony was not well grounded, explanations of the way it worked—that is, of how her first sexual partner "infected" or "saturated" a virgin female—left still more to be desired.

Ewart intended to replicate this long ago series of events at Penicuik, at least as far as he could. The experiment he designed was elaborate, requiring a financial outlay large enough to buy between thirty and forty animals and then to feed and house them.[26] (Doubtless this was one reason why he was eager for results; four years was a relatively short period in which to complete an experiment based on breeding equines.) But expenses and logistics

Quagga—the sire of the first offspring of Lord Morton's mare.

proved to be his least challenging problems. By 1895, when he stocked his farm with the prospective parents of his hybrids, there were no more quaggas to be purchased; they had become extinct rather suddenly several decades earlier. He had to settle for zebras, which were still available on the exotic animal market. Zebras' close relation to the erstwhile quagga and their more prominent coloration made them good substitutes. They also resembled quaggas in being natives of a milder climate, however, and Midlothian did not necessarily suit them. Of the three zebra stallions Ewart acquired, only one—Matopo—survived his first Scottish winter. And physical acclimatization turned out to be only the first step toward procreative success. There were also psychological barriers to overcome. In his initial season among the horse and pony mares, Matopo managed to sire only one foal. But by the next year he and his companions had adjusted to each other, and soon "quite a number of hybrids" made their appearance, although not as many as there might have been.[27] Quite a number of potential hybrids also failed to make their appearance, with Ewart reporting that of four Shetland ponies mated with Matopo, only one produced a hybrid foal; of five Iceland ponies, only one produced a hybrid foal; and of eight full-sized mares (seven thoroughbreds and one Arab), only one produced a hybrid foal. Attempts to cross Matopo with Welsh, Exmoor, New Forest, Norwegian, and Highland ponies were total failures; that is, although mating took place, no offspring resulted.[28]

If Ewart thus made sure that Matopo gave full value for money, he kept his horses and ponies equally hard at work. Once a mare had given birth to a hybrid, she was then mated repeatedly with a stallion of her own breed, to see whether the purebred horse or pony foals that resulted would betray any trace of Matopo. Thus, in effect, Ewart ran several simultaneous replications of the breeding history of Lord Morton's mare, scrutinizing the successive offspring of an Irish mare named Biddy, an Iceland pony named Tundra, a Shetland pony named Nora, and a West Highland pony confusingly named Mulatto.

Nor were the zebra hybrids themselves intended as mere by-products of this research. As a secondary goal of his work, Ewart hoped that they would turn out to be supermules: hardier than their mothers and tamer than their father, and therefore valuable on their own account as pullers of artillery and transporters of supplies in hot imperial locations for which ordinary mules were unsuitable. This subsidiary project looked rather promising at the beginning. One observer admitted that he could "fully confirm all the praise Professor Ewart lavishes on his pets," which were "the most charming and compactly built little animals possible." But although they possessed great stamina and proved generally disease resistant in comparison with full horses and ponies, they were not significantly better able to resist the bite of the tsetse fly. They were nevertheless tested for possible military use by the governments of both India and Germany, but "zebrule" breeding farms on the plains of East Africa remained a fantasy.[29]

By 1899, after only a few breeding cycles, Ewart felt that he had collected sufficient evidence. The skepticism which with he had begun his experiments had been confirmed. Although he would not absolutely deny that "infection of the germ" might occur in unspecified anomalous cases, none of his mares had produced subsequent offspring who resembled Matopo (the "previous sire") in any way. Of Circus Girl, a full-blooded pony borne to Tundra after she had produced two zebra hybrid foals, Ewart stated "there is nothing whatever about her that suggests . . . the zebra"; he went on to assert that "the other half brothers and sisters of the hybrids . . . agree with her in failing to give any support whatever to the . . . telegony doctrine."[30] He confidently attributed the widespread acceptance of "the Mortonian hypothesis" to "the

James Cossar Ewart and a zebra hybrid, from Ewart, *Guide to the Zebra Hybrids*, 1900.

spirit of mediaevalism which is everywhere evident when the application of scientific methods fails to be considered in England."[31] By "spirit of mediaevalism" he meant the instinctive conservatism of agriculturists—their reluctance to replace practices based on their own traditional wisdom with those suggested by modern zoological research.

In order to combat both the particular error and the generally retrograde spirit, Ewart publicized his results as widely as possible. He addressed his elite scientific colleagues in a paper presented to the Royal Society, which included a technical explanation of why telegony could not occur in horses, asses, or zebras. ("All my observations point to its being impossible in the Equidae for the unused male germ cells of the first sire to infect the unripe ova. The spermatozoa lodged in the upper dilated part of the oviduct of the mare are dead, and in process of disintegrating, eight days after insemination; they probably lose their fertilising power in four or five days. There is no reason for supposing that in the Equidae they survive longer in or around the ovary."[32]) He addressed colleagues with more focused interests in articles in the *Zoologist* and the *Veterinarian*, which were subsequently republished in a form that made them available to broader audiences.

His speeches to special interest groups were often refracted through the periodical press. For example, a few months after the RASE show at York, the *Times* carried a detailed account of his keynote address to the National Veterinary Association of Ireland, where the ensuing discussion had focused on the possibility of reversion to ancestral characteristics, especially in the Connemara pony.[33] The next year, as president of the zoological section of the BAAS, he spoke on "The Experimental Study of Variation," using his Penicuik data.[34] He also embraced opportunities to explain his work in person to interested members of the general public. W. Fream, who reported on the York show for the *Journal of the Royal Agricultural Society of England* (devoting about one-fifth of his article to the zebra hybrids, another indication of the society's sense of their significance), praised Ewart's kindness in attending, and noted that he was "unwearying in his efforts to afford to the visitors who inspected the display any information they sought," even if they could have extracted the same information from the *Guide* without much trouble.[35]

But neither the distinctive attributes of Ewart and his animals nor a reciprocal commitment to improved stockbreeding practices completely accounted for the public response to the Penicuik experiments. The widespread interest that they evoked and the relative consistency of presentation from one periodical to another also characterized the treatment of a set of related topics. Hybridity in general was a source of sustained fascination for nineteenth-century audiences. Ewart's half-zebras may have been in the running for hybrid superstardom, along with the litters of lion-tiger cubs that had toured Britain in the 1820s and 1830s as part of Thomas Atkins's menagerie, but almost any hybrid was liable to receive at least brief notice in both general-audience and specialized periodicals. Their newsworthiness was ordinarily taken to be self-evident, so that such reports, wherever they appeared, frequently offered a mere statement of mixture, or even intended mixture, supplemented by a few interesting details about the circumstances (if available). Thus in 1824 the *Annals of Sporting and Fancy Gazette* observed that the Earl of Derby kept "two of those animals of the hog-tribe called the *peccary* . . . for the purpose of trying some experimental crosses"; in 1851 *Notes and Queries* reported that a French "she-wolf" who had been reared with a hound pack "has had and reared a litter of pups by one of the dogs, and does duty in hunting"; in 1888 a correspondent of the *Zoologist* wrote that "it may interest some of your readers if I briefly describe the appearance of some equine Mules which I saw in Paris"; and the *Proceedings of the Zoological Society of London* noted in 1899 that "some living specimens of supposed hybrids between the Stoat *(Mustela erminea)* and the Ferret *(M. Furo)*" had been exhibited at a recent meeting.[36]

The discussion of the Penicuik experiments by Ewart and others clearly demonstrated that the reliable appeal of hybrids also reflected their relation to several large scientific subjects. They provided fodder for debates about the so-called species question, since one traditional, if always problematic, criterion for species difference between two animals was their inability to produce offspring (or fertile offspring, as in the case of everyday horse-donkey mules). This violation of received categories was in fact the source of much of the attractiveness of hybrids, and it continued to be so through the end of the nineteenth century, even though by then, in the view of most zoologists, evolutionary theory had largely mooted the basic problem. Thus Ewart acknowledged the transgressive dimension of hybridity in his *Guide* to the York exhibition by including a detailed illustrated history of equine hybrids of all sorts, even though the rest of his presentation, and, indeed, the design of his experiments, assumed that the issues presented by the zebra hybrids were exactly the same as those presented by crosses between ordinary domestic animal breeds, or, for that matter, by animals whose parents were of indistinguishable heritage.[37] That is, the distinctive striping of the zebra Matopo, like that of Lord Morton's earlier quagga, along with a few other obvious characteristics such as the form of the mane, made it seem relatively easy to distinguish their contribution to their half-horse offspring from that of the mothers, as well as from that of the sires of the mares' subsequent foals. The processes illuminated by the production of these hybrids were not, therefore, especially relevant to interspecies crosses; on the contrary, they were identical to those that determined the outcome of intraspecies and intrabreed matings. The accident of hybridity—and of coloration—made it possible to examine what would otherwise have been obscured by the physical similarity of the actual and putative parents.

Thus Ewart used his zebra hybrids to address general questions of heredity and reproduction, which were of even broader interest, and of much greater practical consequence, than was the species question. His equine research came at the end of a period during which a great deal of zoological attention had been illuminatingly focused on mammalian reproductive physiology. As a result, science could offer answers to some questions that had long perplexed breeders, such as what the contribution of the male and the female parent to the offspring was and (as Ewart both demonstrated and explained) whether it was possible for the first sire to "infect" or "saturate" a female so as to affect her subsequent offspring. (Answers to the related questions why offspring frequently did not resemble either parent and why they sometimes resembled grandparents or still more distant forebears had to wait a little longer.) Like the results of Ewart's work on telegony, these advances in understanding were disseminated to the general public, as well as within the scientific community. They were not, however, readily integrated into the practice, or even the discourse, of those who stood to profit from them most.

Ewart was not alone in his critique of the recalcitrance of the "mediaevalism" of animal breeders. He had a few sympathizers even within this regrettably retrograde group. For example, Everett Millais, a prominent kennel expert who was sometimes known as the father of the English basset hound, shared Ewart's opinion. In his frequent contributions to the kennel press, he bemoaned the failure of his fellow breeders to assimilate the implications of zoological research and to apply them to their own pursuits. He asked, "Why is it that the scientific world accept Darwin's theory, and the unscientific refuse it?" The explanation that he proposed was brutally frank: "It is simple want of power of intellect—a want of education," as was also his characterization of the likely consequences of these failings.[38] He was particularly worried about the extreme inbreeding practiced by some basset hound fanciers, which he considered to violate Darwin's principle of natural selection. If continued indefinitely, he feared, it was likely to result in deterioration of the quality of his favorite breed (that is, the abstraction recognized by show prizes), as well as in unhealthiness and sterility. He predicted that persistent violation of natural laws would invoke an inevitable retribution: "is it likely that Nature . . . will accept such gross liberties with her prerogatives as we breeders take?"[39] But dog fanciers were no more frightened by Millais' threats than stockbreeders were persuaded by Ewart's demonstration of the impossibility of equine telegony. They continued to mate their prize animals within the smallest possible family circles and to guard their maiden females against inappropriate "infection."

In their resistance to scientific expertise, they were, perhaps, playing for different stakes. At the beginning of the Victorian period, naturalists and agriculturalists had faced the mysteries of heredity and reproduction on a roughly equal footing. With little understanding of reproductive physiology and none of the mechanism of heredity, both groups were likely to refer these mysteries to a higher authority. Thus, for example, Thomas Eyton speculated in the *Magazine of Natural History* that since "all true hybrids that have been productive have been produced from species brought from remote countries, and in . . . a state of domestication," it was likely that "it is a provision of Providence, to enable man to improve the breeds of those animals almost necessary to his existence." And John Fry, discussing canine hybrids in the *Hippiatrist and Veterinary Journal,* asked "why should we entertain any doubt that the dog is not a distinct species . . . why question its being formed by the Almighty Framer of the Universe on the sixth day?"[40] Indeed, the fact that the operation of heredity was mysterious, while its effects were ubiquitous and obvious, more than leveled the playing field. The experience and observation of farmers provided at least as firm a basis for speculation as did the experience and observation of naturalists.

Well into the nineteenth century, the speculations of breeders tended to resemble those of naturalists. Both groups were apt to explain the hereditary transmission of characteristics in domestic livestock in terms drawn from other areas of their shared experience. In particular, as was perhaps inevitable when the subject was reproduction, proposed explanations reflected contemporary understandings of human gender relations. Thus the breeders' stubborn belief in telegony, with its corollary imperative of protecting pedigreed virgin females and constraining their choice of partners, bore an obvious relation to Victorian social mores, as did the credulity of scientists in this regard and their own hesitation to jettison the concept completely. Even Ewart left a small loophole, after undertaking elaborate experiments that he regarded as conclusive. Similar preconceptions underlay frequent assertions to the effect that "the influence of the male greatly exceeds that of the female, in communicating qualities to the offspring" and that "the male gives the locomotive and the female the vital organs."[41] Such wisdom appeared most frequently in the agricultural and pet-fancying press, because it was most directly relevant to the pursuits of its readers, but when naturalists addressed these issues their views were often similar. Most famously, with regard to Lord Morton's mare, Charles Darwin was persuaded that "there can be no doubt that the quagga affected the character of the offspring subsequently begot by the black Arabian horse." He further believed that "many similar and well-authenticated facts . . . plainly show . . . the influence of the first male on the progeny subsequently borne by the mother to other males," attributing this phenomenon to some undetermined action of "the male element . . . directly on the female."[42]

By the time Ewart put his hybrids on display, the scientific situation was greatly changed. No longer was it possible to think of the zoological and agricultural discourses of animal breeding and reproduction as parallels or alternatives; scientific research had decisively trumped breeding tradition. The completeness of this triumph reverberated in the smugly confident tone that characterizes both Ewart's and Millais' denunciations of agrarian backwardness. In consequence, no matter how affable his public persona, how sustained his service to agricultural and veterinary causes, or how earnest his desire to improve British cattle, sheep, pigs, and horses, Ewart came to the breeders assembled at York as an emissary from a world of more authoritative expertise. He may have owned a farm at Penicuik, but he was also a Fellow of the Royal Society, and a prominent participant in many of the institutions of elite science. The response to any such powerful ambassador is apt to be ambivalent, part gratification and part resentment.

So why did the visitors to the RASE show at York throng to admire Ewart's exhibit? And why was it featured so prominently in the layout of the show ground and in published accounts? The *Guardian*'s dubious correspondent may have offered a clue. He accorded Ewart and his animals a kind of respect that was simultaneously grudging and skeptical—both Ewart's own professional stature and the conspicuous position accorded the zebra hybrids suggested that the exhibit was important, and yet he could not put his finger on exactly why. His reluctance to commit himself, like the uniformity of reportage on Ewart's work at Penicuik, almost irrespective of which journal did the publishing, and even of whether Ewart or someone else had done the writing, underlined the importance of context in determining the meaning of an exhibition or an experiment or an article. Access to the press, and even strong influence on what was published, did not necessarily imply control of the outcome. The same words could have different implications for an audience of naturalists, for an audience whose major commitment was to agricultural tradition, and for an audience in pursuit of simple amusement.

The planning committee for the RASE show may well have included the exhibition of zebra hybrids as part of an effort to make the fruits of zoological research more accessible to agriculturalists. The application of science to agriculture had, after all, been part of the society's charter since its foundation, and the annual show, "the most generally popular feature of the Society's work," offered the best opportunity to realize this goal—certainly a far better opportunity than that offered by the rather dry *Journal of the Royal Agricultural Society of England*.[43] Ewart's debunking of telegony and his sug-

Understanding Audiences and Misunderstanding Audiences

"Lion-tiger," from Richard Lydekker, *Hand-book to the Carnivora*, 1896.

gestion of alternative principles, more firmly grounded in science, on which to base breeding decisions were also consistent with the more immediately pragmatic purpose of the shows: to allow farmers to see "in what respects their practice or system of breeding is susceptible of improvement."[44]

But the men who ran the RASE lived in a more rarefied atmosphere than did most of the rank-and-file members (or nonmember farmers) who attended the show in order to enjoy themselves for a few early summer days admiring prize animals and large new machines. Millais to the contrary notwithstanding, they were not persuaded that their allegiance to traditional breeding practices had deleterious, or even suboptimal results, with regard to the resulting animals. And such allegiance offered some intangible benefits, bolstering communal self-esteem and providing a kind of passive resistance to the suggestion that they knew less about their own business than did a zoologist whose major experience, when all was said and done, was in the laboratory.

Reports of the exhibition do not suggest that the main impression it made was scientific. On the contrary, the lasting image carried away by show visitors was usually of the hybrid animals themselves, and the same was almost certainly true of those whose experience was mediated by reports in the periodical press. Even the RASE ultimately contributed to this response.

Although Ewart's display received a disproportionate amount of ink in the official account of the show published in the *Journal of the Royal Agricultural Society of England*, most of the coverage consisted of an annotated list (essentially a reproduction of part of Ewart's *Guide*) of some of the items on display, supplemented by photographs of the animals. The article offered no discussion of the critique of telegony. Indeed, its author provided only the vaguest indication of the purpose of Ewart's experiments, merely stating with some evasiveness that they had "a direct bearing . . . upon the many questions that confront the stock breeder" and that Ewart's "explanatory notes . . . convey a clear idea of the problems upon the solution of which he is engaged."[45] Ewart was thus refigured as an impresario surrounded by his exotic creations—a source of wonder and entertainment, but not necessarily of instruction.

Notes

1. For an extended analysis of the financial problems besetting the RASE shows, see "Report of the Special Committee."
2. "Royal Agricultural Society's Show," *Times* (London), 21 June 1900, 14.
3. "The Royal Show at York," *Manchester Guardian*, 19 June 1900, 12.
4. "Royal Agricultural Society's Show at York," *Times*, 16 June, 1900, 16.
5. "The Royal Show at York," *Times*, 25 June 1900, 16. A good turnout would have been more than 120,000. "Report of the Special Committee," 79.
6. Watson, *History of the Royal Agricultural Society*, 64.
7. "Royal Show at York," *Manchester Guardian*.
8. Ibid.; "Royal Agricultural Society's Show."
9. Fream, "York Meeting, 1900," 412–13.
10. Ewart, *Guide to the Zebra Hybrids*, 6.
11. Ibid., 1–8.
12. "Royal Show at York," *Manchester Guardian*.
13. Ewart, *Guide to the Zebra Hybrids*, introductory note.
14. "The Penicuik Zebra Hybrids," *Evening Dispatch*, 4 October 1897.
15. W. B. Tegetmeier, "Zebra Hybrids at the Highland Agricultural Show," *Field*, 15 July 1899, 100.
16. "Excursion to Penicuik," *Scotsman*, 30 July 1898.
17. Cutting from unidentified newspaper, in Ewart, Cutting Book, 1896–97.
18. For example, the *Times*, the *Daily Graphic*, the *Field*, the *Sketch*, the *Live Stock Journal*, and *Land and Water* all carried accounts of the Royal Institution lectures in April and May 1899.
19. "The Zebra Stud Farm at Penicuick," *Polo Magazine*, November 1896.
20. *Sportsman*, 15 January 1898, quoted in Ewart, *Penycuik Experiments*.

21. Quotations from *Times*, 13 March 1899; *Natural Science*, March 1899; *Lancet*, 1 April 1899; *Morning Post*, 5 April 1899; *Scottish Farmer*, 18 March 1899—all appearing on the back cover of Ewart, *Guide to the Zebra Hybrids*.

22. R.F.S., "Penycuik Experiments," *Irish Naturalist* 8 (1899): 116.

23. Arthur Shipley, "Zebras, Horses, and Hybrids," *Quarterly Review* 190 (1899): 422.

24. Natural History Collections of the University of Edinburgh website, http://helios.bto.ed.ac.uk/icapb/collection/.

25. For detailed discussions of telegony, see Ritvo, *Platypus and the Mermaid*, chap. 3; and Burkhardt, "Closing the Door on Lord Morton's Mare."

26. Shipley, "Zebras, Horses, and Hybrids," 422.

27. Ewart, "Experimental Contributions," 248.

28. Ibid., 250–51.

29. Shipley, "Zebras, Horses, and Hybrids," 420–22; "Scientific Notes and News," *Science*, 24 July 1903, 128.

30. Ewart, *Guide to the Zebra Hybrids*, 45.

31. Ewart, *Penycuik Experiments*, 35; idem, *Guide to the Zebra Hybrids*, 50.

32. Ewart, "Experimental Contributions," 245.

33. "Ireland," *Times*, 25 August 1900, 5.

34. J. Playfair McMurrich, "On the Glasgow Meeting of the B.A.A.S.," *Science*, 25 October 1901, 637.

35. Fream, "York Meeting, 1900," 414.

36. T., "A Visit to Knowsley Hall, in Lancashire, the Seat of the Earl of Derby," *Annals of Sporting and Fancy Gazette* 6 (1824): 224; T_____n., "Cross between a Wolf and Hound," *Notes and Queries* 3 (18 January 1851): 39; J. J. Weir, "Equine Mules in Paris," *Zoologist*, 3rd ser., 12 (1888): 102–3; A. H. Cocks, "Hybrid Stoats and Ferrets," *Proceedings of the Zoological Society of London* 67 (1899): 2–3. The *Proceedings of the Zoological Society of London* was probably the most inveterate chronicler of such crosses, noting an unremitting stream of mixed bears, monkeys, cattle, and felines over the years.

37. Ewart, *Guide to the Hybrid Zebras*, 24–35.

38. Millais, *Theory and Practice of Rational Breeding*, ix.

39. Everett Millais, "Basset Bloodhounds: Their Origin, Raison D'Etre and Value," *Dog Owners' Annual for 1897*, 20.

40. Thomas C. Eyton, "Some Remarks upon the Theory of Hybridity," *Magazine of Natural History* 1 (1837): 359; John Fry, "On Factitious or Mule-Bred Animals," *Hippiatrist and Veterinary Journal* 3 (1830): 136.

41. Adam Ferguson, "Some Practical Hints upon Live Stock, in Particular as Regards Crossing," *Quarterly Review of Agriculture* 1 (1828): 34; "The Physiology of Breeding," *Agricultural Magazine, Plough, and Farmers' Journal*, June 1855, 17. For an extended discussion of the influence of human gender relations on animal breeding, see Ritvo, "Animal Connection."

42. Darwin, *Variation of Animals and Plants under Domestication*, 1:435–37.

43. Watson, *History of the Royal Agricultural Society of England*, 18–19; "Report of the Special Committee," 69.
44. "Report of the Special Committee," 72.
45. Fream, "York Meeting, 1900," 414.

— 8 —

Foreword to Charles Darwin, *The Variation of Animals and Plants under Domestication*

Charles Darwin wrote *On the Origin of Species* in a hurry. He had been formulating his ideas and arguments for several decades—since his round-the-world *Beagle* voyage of 1831–36. These ideas and arguments had been slow to take definitive shape; Darwin had nurtured and reworked them, amassing evidence for what he projected to be a weighty magnum opus. Although he had shared his developing evolutionary speculations with his closest professional colleagues, Darwin was reluctant to publish them on several grounds. He was aware that his theory of evolution by natural selection (or descent with modification) was complex, that it rested on vast but not incontrovertible evidence, and that the chain of his reasoning was not uniformly strong. Further, his conclusions challenged not only the scientific assumptions of many fellow specialists but also the theological convictions of a much wider circle of fellow citizens.

In 1859, Darwin did not feel quite ready to expose his cherished theory to the harsh light of public scrutiny. In the introduction to the *Origin* he confessed that although his work on evolution by natural selection was "nearly finished," he would need "two or three more years to complete it." The *Origin* was, he suggested, merely a stopgap, a schematic "abstract" of a much longer and more fully supported treatise yet to come. He had been moved to preview his labors in this way, he explained, because his health was "far

---

This foreword originally appeared in *The Variation of Animals and Plants under Domestication*, vols. 1 and 2 by Charles Darwin (© 1998 The Johns Hopkins University Press), v–xii, and is reprinted with permission of The Johns Hopkins University Press.

from strong" and, perhaps more importantly, because Alfred Russel Wallace, a younger naturalist working in isolation in southeast Asia, had sent a paper to the Linnean Society of London in which he "arrived at almost exactly the same general conclusions that I have on the origin of species." If Darwin had not gone public with his theory at this point, he would have risked losing credit for the work of many years.

As its reception showed immediately and has continued to show, the *Origin* benefited from the succinctness imposed by circumstances. Darwin himself may have appreciated this point; at any rate, he never produced the massive treatise, although he repeatedly issued revised editions of the *Origin*. But he did not abandon his intention to buttress his initial schematic presentation with additional evidence. In the course of the next two decades he published several full-length elaborations of topics summarily discussed in the *Origin: The Variation of Animals and Plants under Domestication; The Descent of Man, and Selection in Relation to Sex;* and *The Expression of the Emotions in Man and Animals*. In addition to fleshing out the *Origin,* these subsequent studies bolstered its arguments and responded to questions raised by critical readers, especially pragmatic questions about the way descent with modification actually operated.

In *The Variation of Animals and Plants under Domestication,* which appeared first in 1868 and then in a revised edition in 1875, Darwin developed a theme to which he had accorded great rhetorical and evidentiary significance. He had begun the *Origin* with a description of artificial selection as practiced by farmers, stockbreeders, and pet fanciers, thus using a reassuringly homely example—one recognizable to the general public as well as to members of the scientific community—to introduce the most innovative component of his evolutionary theory. In addition, domesticated animals and plants, because they were numerous and available for constant observation, provided a readily available body of evidence.

Reassuring as it was, the analogy between natural and artificial selection was far from perfect. The point of Darwin's analogy was to make the idea of natural selection seem plausible by characterizing its efficiency and shaping power. He noted, for example, that some of the prize birds bred by London pigeon fanciers diverged so strikingly in size, plumage, beak shape, flying technique, vocalizations, bone structure, and many other attributes, that if they had been presented to an ornithologist as wild specimens, they would unquestionably have been considered to represent distinct species, perhaps

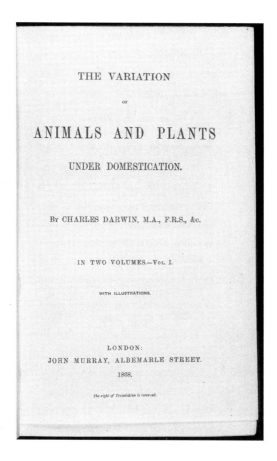

even distinct genera. Darwin argued that if the relatively brief and constrained selective efforts of human breeders had produced such impressive results, it was likely that the more protracted and thoroughgoing efforts of nature would be still more efficacious.

But as Darwin acknowledged, there were some fairly obvious reasons why the two processes might diverge. The superior power of natural selection—"Man can act only on external and visible characters: nature . . . can act on . . . the whole machinery of life. Man selects only for his own good; Nature only for that of the being which she tends" (*Origin*, chap. 5)— might constitute a difference of kind rather than of degree, as might the much greater stretches of time available for natural selection. Further, although the mechanisms of the two processes appeared superficially similar, their outcomes tended to be rather different. Natural selection produced a constantly increasing and diversifying variety of forms; it never reversed or exactly repeated itself. Anyone familiar with artificial selection would have realized that, although new breeds were constantly being developed and although neither improved wheat nor improved cattle showed any tendency to revert to the condition of their aboriginal wild ancestors, the strains produced by human selection were neither as prolific nor as durable as those produced by nature. Indeed, the animals and plants celebrated as the noblest achievements of the breeder's art were especially liable to delicacy and infertility. Highly bred strains, long isolated from others of their species in order to preserve their genealogical purity, far from serving as a springboard for future variation, often had to be revivified with infusions of less rarefied blood. Yet any relaxation of reproductive boundaries threatened subsidence into the common run of conspecifics.

Darwin firmly connected *Variation* to the *Origin* by devoting its introduction to an overview of his theory of evolution by natural selection. In particular, the two volumes of *Variation,* cumbersomely organized and packed with zoological and botanical detail, addressed some of the difficulties inherent in the attractive but paradoxical analogy between natural selection and artificial selection. For selection of any sort to operate, diversity already had to exist. With wild populations living under natural conditions, however, diversity was difficult to discern. It was widely believed that a heightened propensity to vary (at least in ways obvious to human observers) was one of the few general characteristics that differentiated domestic animals as a group from their wild relatives. This point was conventionally illustrated with reference to coat color and design. American bison, for example, were, on the whole, brown, and all Burchell's zebras shared similar black and white stripes. A single herd of either *Bos taurus* (domestic cattle) or *Equus caballus* (domestic horses), on the other hand, could display colors ranging from white through yellow, red, and brown to black, as well as a variety of spotted and blotched patterns.

In order to demonstrate that such populations spontaneously produced sufficient variation to support artificial selection, Darwin devoted most of the first volume of *Variation* to a species-by-species survey of domesticated plants and animals. He began with the dog, the breeds of which differed so greatly in size, shape, disposition, talents, and every other characteristic that Darwin attributed its exemplary plasticity to its derivation from several different species of wild canines. Domestic cats, on the other hand, differed relatively little from one another; at least, their variation tended to be individual, rather than consolidated into breeds. Darwin attributed this to the minimal influence exerted by cat owners over the mating behavior of their animals, so that, alone among fully domesticated animals, cats could not be said to have undergone a genuine process of artificial selection.

Farmyard ungulates, however, had all proved more susceptible to human manipulation, whether through the gradual enhancement of inherent tendencies, such as the relatively early maturation that distinguished shorthorn cattle, or through the preservation of spontaneously arising monstrosities, such as the short, broad foreheads and protruding lower jaws of the niata cattle of South America, the bulldogs of the bovine world. Among animals, fancy pigeons, with their short generations, devoted breeders, and lack of any pragmatic constraints on their extravagant deformations, provided Darwin with his most abundant material. He allotted less space to his survey of domesticated plants, although, with the exception of trees, they tended to be much shorter lived and to be more variable even than pigeons. For example, as Darwin pointed out, a single long-cultivated species—*Brassica oleracea,* the ordinary cabbage—had given rise to strains as distinctive as Brussels sprouts, cauliflower, broccoli, and kohlrabi.

English Fantail (pigeon), from Darwin, *Variation.*

Darwin crammed in so much information of this sort that, in order to confine *Variation* to two volumes of manageable size, less crucial evidence was relegated to a smaller typeface. And so compendious was his survey of domesticates that he felt constrained to deny that it was intended to be an exhaustive catalog. After all, many such catalogs, devoted merely to the accumulation of species- or breed-specific data, existed already; Darwin cited them generously in his footnotes. The material included in *Variation* had been chosen to fulfill a more focused argumentative purpose. Darwin's theory of descent with modification required more than the simple demonstration that abundant variation existed among domesticated animals and plants. The accumulated experience of naturalists and breeders offered no clear explanation of the causes of variation; indeed, no consensus existed on this issue. Variation under domestication was frequently attributed to accidental external influences, especially climate and food. But environmentally induced variation was not of much use to Darwin. Instead, he sought evidence not only that the tendency to vary was inherent in domesticated animals and plants but also that specific variations were inherited.

As a result, Darwin's wealth of detail in *Variation* disproportionately featured strong—as well as puzzling, problematic, or even questionable—versions of inheritance, in addition to the unsurprising, if still not completely understood, likelihood that children would resemble their parents. For example, he devoted an entire chapter to what he termed *atavism* or *reversion*—that is, the tendency for offspring to manifest traits apparently derived from their grandparents, collateral relations, or even remote ancestors, rather than from their mothers or fathers. This tendency in the lineages of individuals, he argued, incontrovertibly demonstrated the fact of heritability; and in an extended or exaggerated version it also demonstrated evolutionary relations among species. Thus, many breeds of domesticated chickens revealed their ultimate ancestry by producing occasional sports with the red and orange plumage of the original *Gallus bankiva*, or jungle fowl.

Like many other naturalists of his time, Darwin was receptive to the idea of telegony, also known as "the influence of the previous sire." He retailed the famous story of Lord Morton's mare, a chestnut of seven-eights Arabian blood, whose first foal had been sired by a quagga (a now-extinct relative of the zebra) that her owner was attempting to domesticate. It was not surprising that the young hybrid faintly echoed his father's stripes, but the fact that her next two foals, both sired by a black Arabian horse, also seemed to resemble the quagga in this regard was more remarkable. Darwin pointed out that atavism offered one possible explanation of this phenomenon—infant horses and donkeys often showed evanescent striping, which might indicate the pattern of their ancient shared progenitor—but he was also drawn to the notion that the first male to impregnate a female left some permanent, heritable trace of himself behind. He offered analogous examples from the vegetable kingdom, where the pollen of related varieties of apples, corn, or orchids not only could produce hybrid offspring but occasionally also physically altered the reproductive tract of the female. Plants also, and more regularly, demonstrated a kind of variability that could arise independently of sexual reproduction, such as "bud variation," whereby what Darwin called a "monstrosity" might appear on a single branch or flower and then be transmitted, sexually or asexually, to future generations.

As he documented the profuse variation among domesticated animals and plants, and the tendency of organisms to transmit these variations down the generations, Darwin did more than demonstrate that there was ample grist for the mill of natural selection. He also addressed the most serious weakness in the argument of the *Origin*. Despite the incompleteness of the fossil record, there was plenty of evidence that evolution had taken place; indeed the idea of evolution had been current in one form or another for a century before 1859. Darwin's explanation of the way natural selection should operate was also widely persuasive. The competitive metaphors with which he characterized it, especially the "struggle for life" prominently featured in the *Origin*'s subtitle, fit well with Victorian understandings about how things worked in the human arenas of industry, commerce, and geopolitics.

There was, however, a problem that troubled not only those inclined to sympathize with Darwin's reasoning but also those inclined to reject it. The efficacy of natural selection, like that of artificial selection, depended on the inheritance of particular traits. But before the modern understanding of genetics, no satisfactory mechanism had been adduced to explain this phenomenon. No consensus yet existed about the way sexual reproduction worked, so there was also disagreement about which characteristics were inherited and which were the result of environment, and about what could be contributed by the male as opposed to the female parent, let alone why offspring sometimes resembled a grandparent or some more distant relative rather than their parents. The special difficulty of accounting for the sudden emergence of monstrosities, or even less dramatically novel traits, led Darwin, in later editions of the *Origin* as well as in *Variation,* to become increasingly receptive to the notion that characteristics acquired by one generation might be inherited by the next.

In the penultimate chapter of *Variation,* Darwin attempted to strengthen the weak link in his chain of argument by proposing a mechanism for inheritance. He called his theory "pangenesis," and he claimed that it explained not only ordinary inheritance—the influence of parents on their children—but also reversion, telegony, the regeneration of amputated limbs in some kinds of animals, the inheritance of acquired characteristics, and the relationship between sexual and asexual modes of reproduction and inheritance. The operation of pangenesis depended on the posited existence of unobservable units that Darwin called "gemmules," tiny granules that were thrown off by individual cells and then circulated through the body. They had, however, an affinity for one another, which led to their aggregation in the reproductive organs or in parthenogenetic buds. They could remain latent for years, until an organism reached a certain stage of development, or for generations, until they encountered other gemmules with which they had some special relationship. In this way a long-dormant great-grandparental gemmule might suddenly manifest itself in a child. Since gemmules could be altered by environmental influences, they could convert acquired characteristics into the stuff of heredity. And since they were vulnerable to error, they could occasionally make mistakes, causing organs, such as limbs or tails or even heads, to develop in inappropriate numbers or in the wrong places.

It has doubtless been fortunate for Darwin's reputation that his theory of pangenesis is not as well remembered as his theory of evolution by natural selection. As vague in detail as it was ambitious and comprehensive in scope, it was unpersuasive at the time and has since been proven completely wrong. But like *Variation* as a whole, which similarly illustrated the limitations of its author as well as his strengths, pangenesis does not therefore lack interest or significance. Despite recent excellent and well-appreciated studies of his entire life and extended *oeuvre,*[1] Darwin is known primarily as the author of the *Origin,* which is unrepresentative in its economy of structure, argument, and evidence, as well as on account of its historical notoriety. Its enforced streamlining has helped to preserve the *Origin*'s accessibility, but its relative paucity of examples was particularly uncharacteristic of Darwin. *Variation,* with its accumulation of evidence about everything from the webbing between dogs' toes to the weight of gooseberries, was much more typical; in addition, it placed Darwin firmly—indeed, irretrievably—within his time, rather than in an achronological limbo reserved for intellectual heroes. As a graduate student from the People's Republic of China told me several decades ago, after participating in a seminar in which he read excerpts from *Variation* and *The Expression of the Emotions*, if the leaders of his government knew that Darwin had written such books, he would not be officially admired.

In science as in politics the victors tend to write the history books. As a result, the record of the past is edited, intentionally or unintentionally, so that it focuses mainly on the precursors of contemporary orthodoxy. Such a focus may accurately represent the genealogy of modern ideas, but it also inevitably misrepresents the historical experience of their progenitors. Viewed without the benefit of hindsight, the marketplace of Victorian ideas seemed much more competitive at the time than it does to us. Even the powerful, persuasive, and ultimately triumphant theory of evolution by natural selection required not only defense, but repeated buttressing and revision. *Variation* showed Darwin hard at work on this rearguard action, using the materials he had at hand—for the most part, homely details about the domesticated animals and plants with which his audience was most familiar. His information was gleaned from the observations of fanciers, breeders, and amateur naturalists, as well as from the treatises of those on the cutting edge of zoology and botany. As hindsight narrows the historical spotlight, it imposes its own sense of hierarchy on the preoccupations of the past. But Darwin was interested in all of these topics, valued all of these sources, and belonged, to a greater or lesser extent, to all these communities.

The author of *Variation* was a Victorian country gentleman, a lover of dogs and horses, a breeder of pigeons and peas. He was also, and equally, the author of *On the Origin of Species*.

Note

1. Browne, *Charles Darwin;* Desmond and Moore, *Darwin*.

Race, Breed, and Myths of Origin

Chillingham Cattle as Ancient Britons

Beginning in the second half of the eighteenth century, British public attention was intermittently captivated by a small but distinguished group of cattle. These striking animals were white (a color not usually favored by British stockbreeders); they were powerfully built; and they roamed the parks of their wealthy proprietors untroubled by the restraints that conditioned the existence of ordinary domestic beasts. At a time when widespread celebration of breeding methods associated with Robert Bakewell emphasized the vulnerability of livestock animals to human manipulation, these cattle gloried in their wildness.[1] The most famous of them lived at Chillingham, the remote Northumberland seat of the earls of Tankerville, and other herds, whose number fluctuated constantly, were scattered across northern England and southern Scotland. Many of these herds were founded during the nineteenth century by landowners who admired the animals. So compelling was their appeal that proprietors who could not afford such a substantial investment in fancy livestock nevertheless occasionally commissioned portraits of their estates adorned by white cattle that, as far as can be determined from any corroborating historical records, never lived there.[2]

Imposing though it was, the physical presence of these animals accounted only in part for their appeal to the British imagination. Their figuration in a variety of discourses—from elite cultural productions to the technical literature of agricultural and natural history to mass-market journalism—

"Race, Breed, and Myths of Origin: Chillingham Cattle as Ancient Britons" originally appeared in *Representations* 39 (summer 1992): 1–22.

suggested that they also carried a serious symbolic charge for a range of eighteenth- and nineteenth-century audiences. That they were, without question, magnificent animals did not sufficiently explain their charisma. Eighteenth- and nineteenth-century Britain was well supplied with animals who might, from one perspective or another, claim magnificence (caged lions, mountainous swine, sheep with eight legs); most of them were lucky to attract attention as simple curiosities. The qualities embodied by (or associated with) certain animals, however, linked them metaphorically or metonymically with issues of great or contentious concern in the human arena. Such connections were powerful, whether or not they were explicit or even manifest to those who made them. The interpretive process that they initiated ran in both directions; once a resonance had been detected or established, it tended reciprocally to determine what people said about the animals in question and even what they did with them.[3] Thus the emparked white cattle, as they were described, pictured, preserved, and admired in their heyday, were complex constructions that both illustrated and helped to shape British notions about such vexed topics as race, descent, and pedigree. That is to say, these bovines by their mere presence—at least as it was understood by many of their human compatriots—addressed questions of origin and identity that were of still more serious concern in the parlor and the meeting room than in the woods and the farmyard.

In 1802, the young Walter Scott edited an anthology entitled *The Minstrelsy of the Scottish Border, Consisting of Historical and Romantic Ballads, Collected in the Southern Counties of Scotland; with a Few of Modern Date, Founded upon Local Tradition*. The collection was a modest commercial success, appealing to the romantic sensibility and sentimental nationalism that had been established in the previous generation by the Ossianic pastiches of James MacPherson and by Thomas Percy's *Reliques of English Poetry;* the entire first edition (only 850 volumes) sold out within a year, and an expanded version appeared in 1803.[4] It contained mostly ballads and other folk poetry featuring the antiquarian and regional motifs that appealed to a growing body of sophisticated readers susceptible to racial nostalgia. But as its subtitle indicated, the collection also contained poems from the editor's own pen, in which such motifs were concentrated or distilled. For example, "Cadyow Castle," which ran to a mere fifty ballad stanzas, packed in castles, hunts, minstrels, fair maidens, marauders, chieftains, English invaders, and religious conflict, as well as plenty of blood.

Predictably enough, most of this blood was human, but some of it came from another source, manifestly also a powerful and evocative symbol of the romantic heritage of the border country. The initial scenes of "Cadyow Castle" focused on a hunt, climaxing in the following lines:

> Through the huge oaks of Evandale
> Whose limbs a thousand years have worn,
> What sullen roar comes down the gale
> And drowns the hunter's pealing horn?
>
> Mightiest of all the beasts of chase
> That roam in woody Caledon,
> Crashing the forest in his race,
> The Mountain Bull comes thundering on.
>
> Fierce on the hunter's quiver'd band
> He rolls his eyes of swarthy glow,
> Spurns with black hoof and horn the sand,
> And tosses high his mane of snow.
>
> Aim'd well the Chieftain's lance has flown—
> Struggling in blood the savage lies;
> His roar is sunk in hollow groan—
> Sound, merry huntsmen! Sound the pryse![5]

The language in which Scott framed this description emphasized the qualities of physical magnificence and ferocious temperament that made the bull both a persuasive incarnation of a landscape presented as almost prehistoric, and an appropriate quarry for a noble huntsman. But it also suggested that their relationship was not circumscribed by such oppositions as victor and vanquished, or pursuer and prey. The attributes that made the bull a desirable trophy—his singularity and preeminence, the fine scorn implied by "spurns" and "tosses," the anthropomorphism of his characterization as "the savage"— also tended to identify him with the heroic chieftain and therefore, ultimately, with the reader.

The dedication that served as the epigraph to "Cadyow Castle" confirmed this set of layered identifications. The poem was "addressed to the Right Honourable Lady Anne Hamilton," a living member of the ancient aristocratic family that had owned Cadzow Castle (as it is more commonly spelled) during the sixteenth century, when the events celebrated in the poem had taken place, and that still owned it in the nineteenth century. Whether or not it had existed at the earlier period, by the time Scott wrote his poem an almost totemic conflation of the Hamiltons with the herd of white cattle that had roamed the Cadzow estate was well established. The power of this legendary association was emphasized by the fact that, at least at the time of Scott's effusion, it apparently persisted without any support from the physical presence of the beasts themselves. He was familiar at first hand with the Hamiltons and their domain, having spent the Christmas of 1801 at Hamilton Palace, the grounds of which contained the ancient park and the ruins of the ancient baronial residence. But he never saw the cattle he described with such admiration. As he put it in his notes to the poem, "There was long preserved in this forest a breed of the Scottish wild cattle, until their ferocity occasioned their being extirpated about forty years ago."[6] That these semilegendary creatures had or came to have significance far beyond the Hamilton family circle may be indicated by the fact that this absence was soon felt to be in need of repair. At least in some versions of the story, the park was restocked with wild cattle in 1809; however, according to other sources, despite the testimony of absence provided by Scott and others, the cattle had never left.

"Cadyow Castle" was only the first occasion upon which white park cattle figured conspicuously in the public culture of nineteenth-century Britain. At the annual Royal Academy show of 1836, the young Edwin Landseer exhibited a striking and much-admired painting alternatively, and perhaps equivalently, entitled *The Death of the Wild Bull* and *Scene in Chillingham Park: Portrait of Lord Ossulston*. The painting recorded a somewhat unplanned occasion. Landseer had been making sketches in preparation for a portrait of the cattle already commissioned by the Earl of Tankerville, and Lord Ossulston, the earl's eldest son, decided to shoot one of the bulls so that Landseer could observe his allegedly distinctive anatomy at closer range than the animal's temperament would otherwise have allowed. Although shooting was the routine method of dispatching the cattle, in this case the procedure went awry, with the result that a keeper was gored and would have been killed if not for the diversionary tactics of Ossulston's deerhound Bran. For his heroism,

The Wild Cattle of Chillingham by Edwin Landseer.

Landseer awarded the dog a place of honor at the left of the painting, which in most other respects was also composed to suggest the conclusion of a hunt rather than the culling of a herd.[7]

Landseer maintained his friendship with the Tankerville family throughout his distinguished career, and he repeatedly returned to Chillingham to sketch their white cattle. Near the end of his life, in 1867, he once again exhibited a painting of them at the Royal Academy. Titled simply *The Wild Cattle of Chillingham,* the portrait featured a dignified family group of bull, cow, and calf posed against a wild landscape and threatening sky. The Lord Ossulston of the earlier painting, long since become the sixth Earl of Tankerville, had commissioned it, along with a companion painting of a stag, doe, and fawn, to adorn the dining hall of Chillingham Castle, where *The Death of the Wild Bull* already hung. In the catalogue of the 1867 Royal Academy exhibition, the entry for

The Wild Cattle of Chillingham included several stanzas from "Cadyow Castle," and taken together, these paintings offered a similarly complex and layered identification of these imposing cattle with both the ancient wilderness they inhabited and the noble family that owned them.[8]

Landseer's selection of these paintings for display at the annual Royal Academy show, as well as the appreciative public response to them, suggested that, at least as far as symbolism was concerned, the white cattle were not the exclusive possession of the aristocratic proprietors with whom they were most closely associated. A happening staged at Chillingham in the autumn of 1872 demonstrated that other members of the aristocratic elite could easily appropriate the animals' cachet. At the invitation of Lord Tankerville (again the sixth earl), the Prince of Wales himself arrived to shoot a bull. Although the confrontation seems to have been fairly well controlled—certainly no "crashing the forest" in the manner of "Cadyow Castle" was allowed, and the prince was well covered in a haycart while "the finest bull in the herd was ridden out" for him to shoot—published reports tended to evoke a wild hunt, with elements of both the medieval chase and the Victorian safari. Writing only a few years after the occasion, a clergyman named John Storer described it as "the great event which of late years has brought the Chillingham cattle prominently before the public . . . the successful pursuit by the Prince of Wales of the noblest unreclaimed animal our country still produces." His narrative of the hunt began with the "right royal welcome" extended to the future monarch, and proceeded through the difficulties of isolating "the king bull" to the inevitable "instant death" of the "noble animal" as the result of a single precisely placed shot.[9] And if the symbolic impact of the white herds was most forceful when they were on the hoof, it did not desert them even when they shared the common fate of cattle. Before the demise of the herd at Lyme Park in Cheshire, each year "one or two animals were shot at Christmas, and some of the beef . . . was always forwarded to Her Majesty the Queen."[10]

Nor were the upper classes alone in their response to the mystique of the white cattle. It was true that in 1848 the popularizing naturalist Philip Gosse claimed that their charisma, and consequently their public audience, had diminished, a diminution that he attributed to the questionable fact that "formerly the hunting of a Bull from these wild herds was attended with much 'pomp and circumstance'; but of late years it has been relinquished from its danger; and now the keeper shoots them as needed." Nevertheless

"Head of a Chillingham Bull Shot by H.R.H. the Prince of Wales," from John Storer, *Wild White Cattle of Great Britain*, 1879.

much evidence suggested that ordinary Britons continued to find a great deal to appreciate and to identify with in these powerful, half-legendary beasts.[11] Indeed, by the last decades of the nineteenth century, the surviving herds of white park cattle had emerged as tourist attractions. One natural-history-oriented traveler enjoyed a visit to the cattle at Chillingham on the spur of the moment, after heavy seas had forced him to cancel his planned trip to the Farne Islands, off the Northumberland coast.[12] At Cadzow, the keeper reported that the cattle had become "much less wild and dangerous . . . in consequence of being visited by so many people."[13] And by 1896, the Earl Ferrers, the hereditary owner of the herd at Chartley in Staffordshire, realized that not only were people eager to admire his animals, but they would actually pay for white cattle memorabilia. Thus, before his annual culling of the herd, he placed an advertisement in the county paper "offering for sale the heads and skins of some Bulls and Cows which were about to be killed.

. . . The applications were so numerous that all the specimens were quickly snapped up at considerable prices."[14]

These striking animals fascinated experts as well as artists and other members of the nonspecialist public, although these parallel attractions were not expressed in identical terms. Shadowing the explicitly symbolic general discourse of literature, art, and popular journalism were several technical discourses that scrutinized the white park cattle through the explicitly empirical lenses of natural history (or zoology) and animal husbandry. Despite their different orientations, all these discourses emerged at about the same time. Until the middle of the eighteenth century, there were relatively few written sources of information about the white cattle, but by the time Scott composed "Cadyow Castle," no survey of the mammalia or of British animals or of domestic livestock was complete without some account of them. And the first quarter of the nineteenth century saw the beginnings of a literature of their own, mostly composed of brief articles but ultimately including several substantial tomes.[15]

The incorporation of the white cattle into the canons of natural history and animal husbandry was decisively confirmed by the official attention of distinguished institutions. The history and taxonomical status of the Chillingham herd were the subject of a report and a lively discussion, both widely reported in the press, at the Newcastle meeting of the British Association for the Advancement of Science in 1838; almost half a century later the same organization appointed a special committee "for the purpose of preparing a Report on the Herds of Wild Cattle in Chartley Park, and other Parks in Great Britain."[16] The British Museum (Natural History) was pleased to receive a sample Chillingham bull from the fifth Earl of Tankerville in 1839 and equally pleased to receive a replacement from his son half a century later.[17] The anatomical collection of the University of Edinburgh contained a single skull from Cadzow, home of the most distinguished Scottish herd.[18] In the 1890s, the staff of the Cambridge University Museum of Zoology went to some pains to accumulate the skeletal remains of bulls from Chillingham, Cadzow ("the second best Bull" in the herd at that time), and Chartley ("the finest Bull in the herd").[19] And at the end of the nineteenth century, a live Chartley bull graced the Regent's Park menagerie of the Zoological Society of London.[20]

The orthodox description of the white cattle took form almost as soon as writers began to describe them. In his *General History of Quadrupeds,* first published in 1790, the engraver and zoological popularizer Thomas Bewick characterized the herds emparked at Chillingham, Chartley, and a few other ancient estates as survivors of "a very singular species of wild cattle in this country, which is now nearly extinct." Relying heavily on the account offered by the eminent agriculturalist George Culley in his *Observations on Livestock* of 1786, Bewick noted that the park cattle were "invariably white," with black or red points; that they sported wide curving horns; and that some of the bulls had manes. Several characteristic behavior patterns seemed to distinguish them from ordinary domestic cattle, and to ally them instead with various wild ungulates: they were apt to form a circle to menace any intruder before charging; cows hid their newborn calves in high grass, returning to nurse them only occasionally; when a member of the herd was seriously wounded or ill, the others turned on it and gored it to death. In the parks where these remnants of the ancient cattle of Britain were preserved, Bewick reported, they were also traditionally treated like wild animals, on analogy with the deer that were much more frequently preserved in this way. That is, they were left to their own devices as far as mating and food were concerned (with such occasional exceptions as the provision of supplemental hay in the hardest winter weather), and they were shot rather than slaughtered, if their continued presence in the herd was considered undesirable or even if they were destined for the table. Bewick characterized "this mode of killing them" as "perhaps, the only modern remains of the grandeur of ancient hunting."[21]

In his *History of Quadrupeds*, published a few years later, Thomas Pennant, a distinguished naturalist and member of the Linnean Society, praised the description offered in what he termed "that little elegant work," and he compounded the compliment by reproducing Bewick's entry almost word for word.[22] Once established, this formulation survived into the Victorian period with relatively little modification. In his introduction to *The Wild White Cattle of Great Britain*, published in 1879, John Storer asserted, "There . . . have existed in this country from the earliest historic times, herds of White Cattle, perfectly distinct, and of a different breed from its ordinary domestic races. Some of these herds seem to have been always wild."[23] His contemporary James Harting, who deployed paleontological as well as historical evidence, claimed that "the few scattered herds of so-called Wild White Cattle which still exist in parks in England and Scotland may be said to form a connecting link . . . between the wild animals which have become extinct in this country within historic times, and those which may still be classed amongst our *ferae naturae*."²⁴ When the Earl Ferrers offered the Chartley cattle for sale in 1904, the advertising circular described them as "the lineal descendants of the original British Wild Ox."²⁵ A few years later a guidebook to Northumberland recommended that tourists visit Chillingham to "see the direct and unmixed descendants of its original wild cattle."²⁶

Indeed, essentially the same formulation has continued to appear in the—admittedly infrequent—modern discussions of these herds. For example, two recent scholarly interpretations of Landseer's Chillingham paintings identified their subjects as "descended from native breeds of wild cattle and ow[ing their] survival to the remoteness of the border country," and as having "roamed the Caledonian forests since pre-Roman times."²⁷ Nor are such views confined to those (such as art historians) who might be presumed to have no grounds for independent judgment. The secretary of the White Park Cattle Society has recently written that "the White Park can claim with confidence to be a truly ancient breed," which is "very distinctive" and "genetically distinct from other British breeds."²⁸

Constant as these reiterated assertions have been, they have not gone unchallenged. Indeed what makes the stability of this formulation over two centuries of technical discussion noteworthy is the fact that every element of it has been repeatedly and persuasively contested. To begin with the most obvious evidences of the singularity of the white cattle, even superficial investigation into the history of the herds suggested that both their color and their horns had always been subject to significant variation, and that their apparent uniformity was more likely the result of sustained human intervention than of long and uninterrupted descent. For example, in his mid-century zoological *Monograph of the Genus Bos,* George Vasey revealed that "it is pretty well known to the farmers about Chillingham (although pains are taken to conceal the fact), that the wild cows in the park not unfrequently drop calves variously spotted."²⁹ More matter-of-factly, the British Association committee of 1887 noted that "a tendency to throw black calves . . . still exists in most of the herds."³⁰

Where nature thus failed to maintain the mystique, artificial selection might step in. Deviations—as both Vasey's exposé and the committee's "Report" suggested, they were rather standard deviations—were summarily dealt with; calves not conforming to the prescribed color pattern were immediately killed. (In most herds, aberrant color provided sufficient justification for this postnatal culling, but sometimes the practice was reinforced by weightier considerations. At Chartley, for example, the slaughter of black calves was attributed to an ancient legend that every such birth portended a death in the proprietary Ferrers family.)[31] Further, it turned out that the ostensibly ancient or primitive standard that determined the fate of newborns was so far from inviolably traditional that it was subject to significant redefinition at the whim of the noble proprietors. Similarly ferocious artificial selection pressures resulted in the elimination of the black points that Bewick recorded in the Chillingham herd; by the late nineteenth century, all the Chillingham cattle had red ears.[32] Even the magnificent lyrate horns of the park cattle turned out to be negotiable. So large a proportion of the minor nineteenth-century herds lacked them completely that Storer divided white cattle into "two pretty distinct varieties," one with horns and one without.[33] But this turned out to be a rather volatile characteristic to use as the basis of varietal distinction. The British Association committee reported on what its members took to be good evidence that the entire Chartley herd had probably been polled a mere forty-five years previously, since "the present duke's grandfather caused all showing the least appearance of being horned to be killed."[34]

If whiteness presented a problem for the owners of the emparked herds, the issue of their wildness turned out to be even more vexed. Enthusiasts loyally echoed the assertion of the fifth Earl of Tankerville that his "wild cattle" were "the ancient breed of the island, inclosed long since within the boundary of the park." For example, Sir Jacob Wilson, writing in a somewhat nostalgic vein at the end of the nineteenth century, noted that "only Chillingham, Hamilton Park and Chartley can truly claim an unbroken pedigree, or that their cattle are still a wild undomesticated race."[35] But increasingly, as time went by and data accumulated, a chorus of demurrals emerged. The underlying question, in the view of skeptics, was whether, even assuming that the nineteenth-century emparked herds lived in a state of nature, an assumption that was at least called into question by the continual pressure of artificial selection, that state represented a historical constant or a relatively modern restoration. Many who investigated the background of the herds concluded that they were feral at best (at wildest, in other words); that is, that however their Victorian lifestyle might be characterized, they were the descendants of animals that had once been completely domesticated.

Some of these demurrals were couched in tentative terms, as was the British Association's diplomatic statement that "whether these animals were

Chillingham wild cattle, from Thomas Bewick, *General History of Quadrupeds,* 1824.

genuine Uri [aboriginal wild cattle], or [merely] feral cattle, admits of some doubt."[36] But many experts asserted with confidence that, wild as the cattle might appear in the nineteenth century, their history must have included, as Richard Lydekker, a late Victorian curator at the British Museum (Natural History) put it, "the intervention of a period of domestication."[37] Different authorities located this domestication at different epochs. For the eminent anatomist Richard Owen it was the Roman occupation of Britain (he claimed to trace the white park cattle to animals imported at this time), while for W. Boyd Dawkins, the paleontologically oriented professor of geology at Owens College, Manchester, it was the medieval period, when, he claimed, all domestic cattle would have seemed wild to Victorian eyes.[38] So confident of their essential tameness was Frances Goodacre, a Norfolk clergyman who had dedicated himself to a lonely campaign for the recognition of a discipline that he called "hemerozoology," or the study of domestic animals, that he used the Chillingham cattle to mark one end of the scale of domestication, the other terminus of which was represented by the dog.[39]

Queries about the wildness of the white cattle inevitably cast doubt on their claims to indigenousness or aboriginality, since all such claims were ultimately based on a narrative that identified the emparked herds of the nineteenth century with animals that had roamed the prehistoric forests of Britain. It was widely known and occasionally regretted that the pressure of human settlement and human hostility had contributed to the extinction of several species of mammals in medieval and even postmedieval times—the wolf, the wild boar, the bear, and the beaver—and those who subscribed to this account viewed the white cattle as an analogous remnant of the aurochs, or wild ox.[40] They had survived, it was argued, as a result of a fortunate conjunction of circumstances: first, their location in remote, relatively unpopulated areas, and second, their protective enclosure by medieval, or at least long-dead, magnates. The element of protection was especially important because, even after emparkment, according to many chroniclers, the continued existence of the white cattle was threatened repeatedly and from varied directions. For example, in the first known written record of the Chillingham herd, which dates from the upheavals of the 1640s, one of the estate servants complained that occupying Scots troops were slaughtering the cattle.[41]

Nor did the public attention the cattle enjoyed in the nineteenth century necessarily signal the end of their perils. During the 1860s, the Cadzow herd, along with all other British bovines, was endangered both by rinderpest, or cattle plague, and by the policy of quarantine and slaughter promulgated by authorities to contain it. Nature determined that only about half the animals succumbed to the disease; the keepers protected the survivors from the law by hiding them in abandoned mine pits until the epizootic had passed.[42] And indeed, although keepers saved the cattle in this instance, for many of the smaller emparked herds preservation turned out to be a double-edged sword, especially when it was accompanied by a zealous interest in genealogical purity. In the course of the nineteenth century, according to a definitive twentieth-century estimate, more than a dozen herds disappeared as the result of the breeding and husbandry policies pursued by their owners.[43]

In the view of many investigators, however, it was far from clear exactly what was being preserved by these noble efforts. The claim that the ancestors of the white cattle were quintessentially British was countered by a range of alternatives. Most dissenting authorities supported Richard Owen's suggestion that the ancestral white cattle had been introduced to Great Britain by the Romans. R. Hedger Wallace offered a concise and forthright version of this position when he disparaged "a commonly accepted view regarding white cattle . . . that they are the true descendants, in an unbroken line, of the aboriginal cattle that existed in Britain"; instead, he asserted, "these cattle are simply the descendants of Roman cattle."[44] But virtually every human migration into Britain might be regarded as a possible source of the white cattle. Thus the anonymous author of an article entitled "Origin and Early Progress of Our Breeds of Polled Cattle" derived them from the cattle of the Celts, while W. Boyd Dawkins identified the herds that accompanied the Germanic invaders as their ancestors.[45]

These historical speculations were buttressed by evidence provided by the cattle themselves. That is, the surviving white herds of nineteenth-century England and Scotland did not resemble one another as closely as might have been predicted on the basis of their postulated common and exclusive descent. Simple observation revealed that some had impressive and terrifying horns, while others were polled. Some had black points, while others had red. The tendency to depart from the white-with-colored-points patterns was widespread, but different herds were apt to depart in different directions. And proprietors were prone to exaggerate the differences between the herds they owned and their alleged relatives, ordinarily with the implication that not all those who claimed white cattle status really deserved it. Thus every historian of the small herd at Lyme Park, which died out late in the nineteenth century due to increasing infertility, at least in part as the result of successive proprietors' vigilance in maintaining the unsullied purity of their animals' lineage, repeated the monitory tale of the black calf born as the result of an attempt to reinvigorate the herd by introducing a bull from Chartley. Previously, it was said, there had been "no record of any departure from the legitimate white ground-colour."[46]

Nor were such inimical comparisons confined to superficial characteristics. In the course of describing his own noble animals, the fifth Earl of Tankerville offered a sweeping disparagement of the Cadzow herd: "They in no degree resemble those at Chillingham. They have no beauty, no marks of high breeding, no wild habits, being kept . . . in a sort of paddock." (It must be said that Tankerville was not uniformly ungenerous. He did think that "those at Chartley Park . . . closely resemble ours in every particular . . . except some small difference in the colour of their ears." In any case, such aficionados of the Cadzow herd as the local clergyman immediately leapt to its defense, castigating Tankerville's account as "full of blunders.")[47]

And if the emparked herds did not resemble one another, they were often perceived to resemble cattle with no high claims to ancient and distinct ancestry. As David Low, the professor of agriculture at the University of Edinburgh, asserted, "no real distinction exists between the Wild Oxen of the parks, and those which have for ages been subjected to domestication."[48] In particular, emparked herds might seem like white versions of one or another of the ordinary domestic breeds, usually a strain common in or near their locality. Thus the Chillinghams were likened to the Ayrshires and Kyloes of Scotland, the Chartleys to the longhorns once plentiful in the Midlands, and the "half-wild" white cattle of Dynevor Castle in Carmarthenshire to the old Castlemartin breed of South Wales.[49]

In addition, despite the claims of the keepers and proprietors of the white cattle that each herd comprised an isolated breeding unit, local rumors often fueled the suspicions that had been sparked by the cattle's appearance. One or two Highland bulls were reported to have bred in the Cadzow herd "some years ago," according to the British Association committee of 1887.[50] Even at Chillingham, barriers between the wild herd and neighboring domesticates were not so strictly defended before the nineteenth century as they would be subsequently. It was well known that local cows had been admitted freely to profit from the attentions of the reigning white bull. Of course, such donations would not have altered the inheritance of the park cattle, but it was also considered likely that in less punctilious times alien bulls had repeatedly been allowed, or even encouraged, to make reciprocal donations to the cows of the white herd.[51]

Nineteenth-century discourses about British white cattle, both popular and technical, were thus structured by a series of internal contradictions. Every assertion was made with great confidence, and as energetically countered and denied. The ever-increasing documentation (at least when counted in terms of pages or pounds of published material) derived from a body of evidence that was small to begin with and did not grow quickly. Before their late-eighteenth-century emergence into the public eye, not much had been written about the cattle. Authors looking for ancient authority had to be satisfied with the brief description of Caledonian wild cattle offered by Hector Boece or Boethius in his early-sixteenth-century *Scoticorum Historiae*, or with occasional (and usually ambiguous) passing references to wild cattle in manorial records and other medieval accounts. For example, in 1839 a contributor to the *Annals and Magazine of Natural History* proudly reported that he had discovered a reference to "tauri sylvestres" in a Latin history of the monastery of St. Albans, which proved that, at least at the time of Edward the Confessor, wild cattle existed "not only in the forests of Caledonia and north of England, but in the midlands."[52]

Even where the existence of white herds in the eighteenth and nineteenth centuries seemed to guarantee their historical presence, and their traditional association with noble proprietors might have been supposed conducive to record keeping, details were difficult to come by. Although, as Charles Darwin noted when discussing the cattle in *The Variation of Animals and Plants under Domestication,* a thirteenth-century document testified to the existence of the park at Chillingham, no written evidence about its renowned bovine inhabitants surfaced until almost half a millennium later.[53] As the fifth Earl of Tankerville noted in 1839,

> All that we know and believe . . . rests in great measure on conjecture, supported, however, by certain facts and reasonings, which lead us to believe in their ancient origin, not so much from any direct evidence, as from the improbability of any hypothesis ascribing to them a more *recent* date. I remember an old gardener of the name of Moscrop, who died about thirty years ago, at the age of perhaps eighty, who used to tell of what his father had told him as happening to him, when a boy, relative to these wild cattle; which were then spoken of as wild cattle, and with the same sort of curiosity as exists with regard to them at the present day.[54]

The chronicles of other Victorian herds reiterated this reliance on the oral testimony of retainers, frequently aged or even deceased. Writing about the Lyme Park herd in 1891, Charles Oldham regretted being unable to speak to "John Sigley, the old keeper, who would perhaps have been able to give me more information than anyone else" but who had unfortunately been dead an imprecise "five or six years"; instead he had to be satisfied with "old Jim Arden, who has been at Lyme, man and boy, for seventy years or more."[55] Investigating a recently extinguished herd in Kirkcudbrightshire at about the same time, Robert Service cited as authority for its size "some of the old people" (their estimates apparently varied by a factor of two); for additional local evidence he relied on a pseudonymous and retrospective letter about a different herd formerly at Ardrossan, in Ayrshire, that had been published in a local newspaper in 1817.[56] And as the Victorian inclination to doubt Sir Walter Scott's eyewitness assertion that the Cadzow herd was absent from its traditional haunts in 1801 indicated, the very existence of one of the most important and allegedly ancient herds for a fifty-year period beginning in the middle of the eighteenth century was the subject of repeated disagreement among a later generation of experts.

Indeed, the information explosion that transformed so many fields during the nineteenth century left the white park cattle relatively untouched. And despite great curiosity, their present condition often seemed as mysterious as their history and origins. Expert observation of the living herds was not much more productive than scholarly scrutiny of the sparse documentary record. The fact that they were allowed to roam freely over their wild domains made it difficult to study their behavior; one data-starved naturalist suggested in 1865 that, by way of repairing this gap, "some valuable information might be collected were any readers of the 'Zoologist' who live near such cattle to favour us with notes on their habits."[57] Their reclusive and independent lifestyle also precluded any systematic (or at least completely reliable) attention to their reproductive behavior, and therefore to their pedigrees, although, inspired by Darwin, the sixth Earl of Tankerville did keep nongenealogical demographic records of births, deaths, numbers, and "remarkable occurrences" in the Chillingham herd from 1862 to 1899.[58] But in general, credible information was hard to come by. As was lamented, somewhat paradoxically, in the founding volume of the *Park Cattle Society's Herd Book* in 1919, "although the different herds have been from time immemorial jealously guarded from any outcross, the names, ages, and pedigree of individuals have often been insufficiently kept or not kept at all."[59]

Thus, at the center of the garrulous public discussion of the emparked herds was a virtual vacuum, even a black hole. And although the absence of reliable data was regretted by some experts, the implicit testimony of others suggested that even this dark cloud had a silver lining. The dearth of information could license the subordination of evidence to the less concrete considerations that in any case characterized the interpretation of these animals. If most of what seemed to be known about them was ambiguous and insubstantial, then the flamboyant white cattle could function as a kind of tabula rasa, upon which people could inscribe their own concerns. For the most part, these concerns derived from the identification of these allegedly primordial cattle with the modern human inhabitants of their island; they were, as one nostalgic writer put it, "ancient Britons."[60] Not only were the surviving herds the totems of distinguished and powerful families, but they inspired considerable local pride. Thus much humbler neighbors of the Tankervilles could state in the *Transactions of the Tyneside Naturalists Field Club* that "our local mammalian fauna is honorably distinguished from that of nearly every other part of the kingdom by the possession of this noble animal in a state of wildness."[61] And as the qualities represented by the white cattle were attractive to all Britons, the metonymic chain could easily be extended further, to the nation as a whole.

The most obvious of these qualities was commanding physical magnificence, emphasized both in visual representations (which in celebrating the animals' physiques seldom hinted—what was also true—that they tended to be smaller than ordinary domestic cattle) and in the almost obsessively reiterated accounts of the difficulty of hunting them.[62] (The iconoclastic Vasey recognized the mythopoeic tendency of these accounts, commenting that the stereotypical scene—for example, the one depicted in Landseer's picture—"does not appear to have much relation to the history of the *Genus Bos*: it however exhibits the brutal and ferocious habits of two varieties of *Genus Homo*, namely *No*bility and *Mo*bility.")[63] The males, and especially the alleged leaders of herds, often referred to as the "monarch" or the "king bull," were the most frequent objects of artists, anatomists, and museum curators, and they were the exclusive objects of hunters. The internal organization ascribed to the herds reinforced the sense of power and domination symbolized by the mere presence of such impressive chiefs. Each herd was conceived as the harem of a single bull, who had to defend his position against continual challenges from other males eager to replace him. (In order to reduce the damage alleged to result from such challenges, in many herds keepers castrated a large percentage of male calves soon after birth, thus also producing another significant, if undirected, artificial influence on the animals' pedigrees.) And these exemplary cattle were also perceived to display the gender stereotypes consistent with their paternalistic polity. According to one admiring observer of the Chillingham herd, "grand, masculine, and majestic is the bull; peculiarly sweet, feminine, and elegant is the cow."[64]

The metaphoric appeal of ancient genetic isolation may have been still more attractive to citizens of the island nation than that of brute strength and firm fatherly guidance. Not every enthusiastic account of the white cattle mentioned their macho social organization, but all celebrated the noble and exclusive descent that justified their separation from the inferior circumambient strains of domestic cattle. So important was this component of the white herds' reputation that the rumors of miscegenation occasionally attaching to one herd or another tended not to be disdainfully ignored but rather energetically denied. Thus it was claimed that at Chillingham "the introduction of alien blood has been rigidly prohibited"; a defender of the Vaynol Park herd in Caernarvonshire earnestly explained that "it was stated a short time ago that these had been crossed with the Pembroke White Cattle. This is a mistake. Some were purchased a few years ago, but none were ever put near the main herd, Mr. Assheton-Smith being most particular in this respect."[65] Indeed, because centuries-long breeding within very small reproductive communities—the largest of the white herds contained fewer than one hundred individuals, and most were much smaller—seemed to contradict a great deal of evidence accumulated by breeders of ordinary livestock illustrating the dangers of extended inbreeding, it was occasionally asserted by dissenters from the general wisdom that the continued vigor of, for example, the Chillingham herd provided "one of the most conclusive arguments that crossing with different stock is not necessary."[66]

If the special status of the white park cattle was thus explained and guaranteed by their unsullied descent, it was reified through their classification. In the second half of the nineteenth century, some strains of human racial theory came to focus on increasingly subtle differences as a means of demonstrating increasingly slippery superiorities. Thus Robert Knox, an influential exponent of such racially based analysis, criticized his predecessors Johann Blumenbach and J. C. Prichard for concentrating too much on the large differences between, for example, Europeans and Asiatics, and too little on the gaps that separated what he referred to as "the European races." To differentiate the Saxons from their nearest neighbors, the Celts, he invoked a variety of criteria. He praised the Saxons physically, describing them as "powerful and athletic . . . the only absolutely fair race [in color.]" But the most telling comparison was on the intellectual level. For example, he attributed the orderly distinction of the Musée d'histoire naturelle in Paris when it was headed by Georges Cuvier to his essential Germanness, and claimed that after Cuvier's death "fanaticism and prudery had been at work" in the collection, resulting in its relapse into its previous condition of Celtic chaos.[67] Insistence on similarly elusive distinctions could be the means of achieving for their metonymic cattle the same glorification through taxonomy enjoyed by human Britons. Thus, although there were wide differences of opinion about exactly how these animals should be classified, their enthusiastic admirers agreed that it was essential to place them in a separate category from ordinary domestic cattle, with their humble origin and low destiny.[68]

In describing a discourse that, though laden with symbolic meaning and fraught with emotion, concerned zoological objects manifestly similar, if not necessarily identical, to animals about which a great deal was positively known and had been known for a long time, one might be tempted to differentiate a sentimental lay position from a more rational expert position. But reading the technical and popular discourses side by side does not encourage such a distinction. In the first place, the dichotomy between the technical and the popular is often difficult to maintain, especially with regard to writers whose expertise lay in animal husbandry or natural history, rather than explicitly in zoology.

And in any case—recurring to this problematic dichotomy—the entire range of opinion was represented within both discourses. It was not, of course, represented in the same terms; specialist literature invariably deployed its characteristic vocabulary and often recast the issues so that they dovetailed with or contributed to other contemporary debates. Thus, when popular authors discussed whether or not the white cattle were primeval or aboriginal, the experts focused on whether they had descended from *Bos longifrons,* then considered to be the ancestor of all other British domestic strains, or from the more exciting *Bos primigenius,* or aurochs.[69] When popular authors wondered about the purity of their lineage, the experts asked whether a modern type might be descended from more than one specific ancestor. And when popular authors argued about the relationship (or lack thereof) between the park cattle and various domestic breeds, the experts debated their taxonomic status: Were they merely a variety of *Bos taurus,* a separate species (often denominated *Bos scoticus*), or perhaps even an independent genus (usually designated *Urus*)?[70]

Nevertheless it was also true that, as time went by, an increasing proportion of those who seemed to know best inclined to what has become the modern scientific consensus about these cattle. Buttressed by the results of recent research, it tends toward skepticism about the more romantic claims of the eighteenth and nineteenth centuries. Thus osteological analysis has

Chillingham bull.

established that the skeletons of modern Chillingham cattle strongly resemble those of medieval domestic cattle. Blood grouping studies have shown that the herd is genetically uniform, which might confirm traditional claims about its reproductive isolation or might simply reflect the genetic bottleneck that occurred after the severe winter of 1947, when its population crashed from approximately forty to thirteen animals. Ethological observation has suggested that the dominance relations among the cattle are far more complex than the clear-cut patriarchy evoked by the term *king bull*.[71] Recognizing the interpretive limits of such evidence, however, this consensus also tends toward resignation about the likelihood of answering some traditionally important questions, such as the ultimate origin and previous condition of servitude of the white herds.

But not all postmodern authorities are so careful about where they tread. A recent contributor to the *Ark*, the journal of the Rare Breeds Survival Trust—in fact, the president of the Chillingham Wild Cattle Association, Ltd.—asserted confidently (in an attempt to gainsay a previous contributor's conflicting assertion) that "although there is still much that is not known about the origins of the Chillingham Wild Cattle, one fact that is certain is that they were never domesticated."[72] And in a pamphlet available to visitors

Race, Breed, and Myths of Origin

to the Wild Cattle Park at Chillingham (the Chillingham estate was sold on the death of the ninth Earl of Tankerville in 1980, at which time a charitable trust purchased the park, ensuring that the cattle could continue to roam their accustomed haunts), the Dowager Countess of Tankerville described the cattle as "direct descendants of the original ox which roamed these islands before the dawn of history"; she explained that because "the fittest and strongest bull becomes 'King' . . . Nature seems . . . to have ensured the carrying forward of only the best available blood."[73]

Nor are Chillingham aficionados alone in their inclination to repeat the certainties and reify the categories of a hundred years ago. The relationship between the various white herds that troubled the Victorians has been sorted out by institutionalization into two breed societies, roughly along the lines suggested by Storer—that is, White Parks have horns and British Whites have no horns. It is probably too early to speculate about what is at stake in these contemporary constructions or reconstructions, but it as least suggestive that although the seventh Earl of Tankerville was the patron of the Park Cattle Society when it was founded in 1918, the Chillingham herd was withdrawn from its herd book in 1932 and now goes its own more nearly feral, if not wild, way.[74]

Notes

1. Russell, *Like Engend'ring Like*, chronicles this development. Bakewell's methods were widely, but far from universally, celebrated. For a discussion of the resulting controversy, see "Possessing Mother Nature: Genetic Capital in Eighteenth-Century Britain," chapter 10 in this volume.

2. Hall and Clutton-Brock, *Two Hundred Years of British Farm Livestock*, 42; Whitehead, *Ancient White Cattle*, 101–50.

3. For a more elaborate version of this argument, see Ritvo, *Animal Estate*, 3–6.

4. Quayle, *Ruin of Sir Walter Scott*, 23–25.

5. In a note, Scott defined the "pryse" as "the note blown at the death of the game." Scott, *Poetical Works*, 690.

6. McCabe, "White Cattle of Cadzow"; Edgar Johnson, *Sir Walter Scott*, 1:186; Scott, notes to "Cadyow Castle," in *Poetical Works*, 689.

7. Ormond, *Sir Edwin Landseer*, 126–27; Lennie, *Landseer*, 221. This story was inevitably rehearsed by Victorian chroniclers of the Chillingham herd, including the fifth Earl of Tankerville himself, although he did not bother to mention Landseer, whose need for a carcass had triggered the episode, in his account. See Tankerville and Hindmarsh, "On the Wild Cattle of Chillingham Park," *Athenaeum* 565 (25 August 1838): 611.

8. Ormond, *Sir Edwin Landseer*, 210–12; Stephens, *Memoirs of Sir Edwin Landseer*, 108.

9. Alfred Heneage Cocks, "A Visit to the Existing Herds of British White Wild Cattle," *Zoologist* 2 (1878): 275; Storer, *Wild White Cattle of Great Britain*, 165–67.

10. Charles Oldham, "The Lyme Park Herd of Wild White Cattle," *Zoologist*, 3rd ser., 15 (1891): 85.

11. Gosse, *Natural History*, 202–3.

12. T. H. Nelson, "A Visit to Chillingham Park," *Naturalist*, August 1887, 229–34.

13. Bidwell et al., "Report of the Committee," 138.

14. J. R. B. Masefield, "Zoology," *North Staffordshire Field Club Annual Report and Transactions* 31 (1896–97): 48.

15. All of the tomes and a representative selection of the rest of the literature are referred to elsewhere in these notes.

16. Accounts appear in the *Report of the British Association for the Advancement of Science* 8 (1839) and 57 (1888).

17. Oldfield Thomas, "Mammals." In *History of the Collections*, 2:58.

18. R. C. Auld, "The Wild Cattle of Great Britain," *American Naturalist* 22 (1888): 508–9.

19. Cambridge University Museum of Zoology, "History Index," items 252 and 253; idem, "Additions to the Museum," 181–82, 221–22.

20. McCabe, "White Cattle of Cadzow," 246.

21. Bewick, *General History of Quadrupeds* (1822), 38–41.

22. Pennant, *History of Quadrupeds*, 1:17–20.

23. Storer, *Wild White Cattle of Great Britain*, xv.

24. Harting, *British Animals Extinct Within Historic Times*, 213.

25. Tavinor, "Chapter in the History of the 'Chartleys,'" 379.

26. Bradley, *Romance of Northumberland*, 86.

27. Ormond, *Sir Edwin Landseer*, 126; Lennie, *Landseer*, 221.

28. G. L. H. Alderson, "History, Development, and Qualities of White Park Cattle."

29. Vasey, *Monograph of the Genus Bos*, 143.

30. Bidwell et al., "Report of the Committee," 135. This tendency persists in surviving herds. The Rare Breeds Survival Trust, which owns the last of the Vaynol cattle, decided not to cull black calves because the numbers were so low. As a result, the herd has become three-quarters black. Juliet Clutton-Brock, pers. comm.

31. J. R. B. Masefield, "The Wild Cattle of Chartley, Staffordshire," *Nature Notes* 9 (1898): 46.

32. Harting, *British Animals Extinct Within Historic Times*, 235.

33. Storer, *Wild White Cattle of Great Britain*, 358–59.

34. Bidwell et al., "Report of the Committee," 137.

35. Tankerville and Hindmarsh, "On the Wild Cattle of Chillingham Park," 611; Jacob Wilson, "The Chillingham Wild Cattle," *Land Magazine*, January 1899, 14.

36. Bidwell et al., "Report of the Committee," 135.

37. Lydekker, *Hand-Book to the British Mammalia*, 237.

38. Richard Owen, "On the Ruminant Quadrupeds and the Aboriginal Cattle of Great Britain," *Proceedings of the Royal Institution of Great Britain* (1856): 4 (offprint); W. Boyd Dawkins, "The Chartley White Cattle," *North Staffordshire Field Club Annual Report and Transactions* 33 (1898–99): 49.

39. Goodacre, *Few Remarks on Hemerozoology*, 6.

40. This analogy explains Harting's inclusion of a chapter entitled "Wild White Cattle" in *British Animals Extinct Within Historic Times*.

41. Dodds, *History of Northumberland*, 301.

42. Page, *Victoria Country History of the Counties of England: Staffordshire*, 167; Edward R. Alston, "Wild White Cattle," *Zoologist*, 2nd ser., 1 (1866): 242, 512.

43. See "Descriptive List of the Ancient and Modern Herds of White Cattle in Britain," in Whitehead, *Ancient White Cattle*, 101–50.

44. R. Hedger Wallace, "White Cattle: An Inquiry into Their Origin and History," *Transactions of the Natural History Society of Glasgow* 26 (1900): 220, 222.

45. "Origin and Early Progress of Our Breeds of Polled Cattle," *Banffshire Journal* (1881): 3 (offprint); Dawkins, "Chartley White Cattle," 50, 54.

46. Oldham, "Lyme Park Herd," 85–86.

47. Tankerville and Hindmarsh, "On the Wild Cattle of Chillingham Park," 611; William Patrick, "On the Ox Tribe, in Connexion with the White Cattle of the Hamilton and Chillingham Breeds," *Journal of Agriculture* 9 (1839): 528.

48. Low, *Breeds of the Domestic Animals*, 1:4.

49. Vasey, *Monograph of the Genus Bos*, 150–51; Whitehead, *Ancient White Cattle*, 19–20.

50. Bidwell et al., "Report of the Committee," 137.

51. Storer, *Wild White Cattle of Great Britain*, 213, 215.

52. Matthew Paris, *Vitae Sancti Albani Abbatum*, quoted in R.T., "On the Wild Cattle of Great Britain," *Annals and Magazine of Natural History* 3 (1839): 356.

53. Darwin, *Variation of Animals and Plants under Domestication*, 1:107.

54. Tankerville and Hindmarsh, "On the Wild Cattle of Chillingham Park," 611.

55. Oldham, "Lyme Park Herd," 83.

56. Robert Service, "Wild White Cattle in South-western Scotland," *Zoologist* 11 (1887): 449–50.

57. Edward R. Alston, "Notes on the Wild Cattle at Cadzow," *Zoologist* 23 (1865): 9514.

58. Hall and Clutton-Brock, *Two Hundred Years of British Farm Livestock*, 45.

59. Claud Alexander, introduction, *Park Cattle Society's Herd Book* 1 (1919): 9.

60. Bradley, *Romance of Northumberland*, 89.

61. H.T. Mennell and V. R. Perkins, "Wild Cattle of Chillingham," *Transactions of the Tyneside Naturalists Field Club* 6 (1863–64): 140.

62. In representing these relatively small-bodied cattle as large, artists were abetted by the animals' unusual color and by the fact that, unlike many domestic cattle breeds, they were (and are) not disproportionately short-legged. Stephen J. G. Hall, pers. comm.

63. Vasey, *Monograph of the Genus Bos,* 146.

64. Jacob Wilson, "Chillingham Wild Cattle," 23.

65. Ibid., 22; Laisters F. Lort, "The White Cattle of Vaynol Park," *North Staffordshire Field Club Annual Report and Transactions* 33 (1898–99): 55.

66. Bates and Bell, *History of Improved Short-horn or Durham Cattle,* 8.

67. Knox, *Races of Men,* 44–45, 50; idem, *Great Artists and Great Anatomists,* 18–19, 80.

68. Knox tended to be skeptical of claims on behalf of the white cattle's ancient, pure, and wild descent, in part as a result of a different analogy between the human and bovine inhabitants of Britain. He asserted that the most primitive cattle of Britain should be found in association with the most primitive people, whom he located in Cornwall, where the local cattle were black. [Robert] Knox, "Anatomical Examination of the Wild Ox of Scotland, with Some Remarks on Its Natural History," *Quarterly Journal of Agriculture* 9 (1838): 383–84.

69. According to subsequent scholarship, the bones identified as *Bos longifrons* in the nineteenth century do not represent a separate wild species ancestral to domestic cattle, or *Bos taurus,* but merely the small domestic cattle of Neolithic Europeans. Along with all modern breeds, it is now considered to descend from *Bos primigenius.* Clutton-Brock, *Natural History of Domesticated Animals,* 64–65.

70. Representative taxonomic ruminations can be found in Patrick, "On the Ox Tribe," 531–34; and in the anonymous "British Wild Cattle," *Penny Magazine* 7 (1838): 443.

71. The most comprehensive work on British white cattle is still Whitehead's *Ancient White Cattle.* For summaries of recent research, see Hall and Clutton-Brock, *Two Hundred Years of British Farm Livestock;* Hall, "White Herd of Chillingham"; and idem, "Running Wild," 12–15, 46–49.

72. Bennet, "Chillingham Cattle," 22.

73. Dowager Countess of Tankerville, *The Wild White Cattle of Chillingham* (N.p., n.d.), 1–4.

74. Hall and Clutton-Brock, *Two Hundred Years of British Farm Livestock,* 42.

— 10 —

Possessing Mother Nature

Genetic Capital in Eighteenth-Century Britain

Once upon a time, there was a man named Robert Bakewell (1725–95), who lived on a large farm called Dishley Grange in Leicestershire, admired by his neighbors, respected by his employees, and beloved by his animals, to whom he was unusually kind. He never married, but devoted himself entirely to livestock husbandry. As a result of years of selfless and patriotic dedication, he presented to his countrymen, who demanded increasing quantities of fresh meat as a result of their burgeoning numbers and intermittent wars with the French, improved strains of the most important domestic species—more succulent sheep and cattle, larger carthorses, and pigs that his friends, at least, described as "superior."[1] One of these improved strains, the New Leicester or Dishley sheep, appeared to be of such transcendent merit that it dominated British sheep-breeding for decades. The production of these distinguished creatures was not the result of lucky accident, nor even of the instinctive application of craft expertise, although Bakewell had plenty of that. More important, however, were his general ideas about how to select superior animals and then pair them so as to ensure that their desirable qualities would predictably reemerge in their offspring and more remote descendants. Based on repeated inbreeding, Bakewell's method had still greater impact on his fellow agriculturalists than did the animals who exemplified it; together, his precepts and his example laid the foundation for

---

"Possessing Mother Nature: Genetic Capital in Eighteenth-Century Britain" originally appeared in *Early Modern Conceptions of Property,* edited by John Brewer and Susan Staves (Routledge, 1994), 413–26.

the British preeminence in stockbreeding that lasted through the nineteenth century.

Bakewell's lifetime of service and achievement made him one of the patron saints, or at least one of the presiding geniuses, of the agricultural revolution. Like Newton and his apple or Franklin and his kite, albeit with a more restricted audience, the Leicestershire improver and his fat sheep became icons of the Enlightenment. One of his early disciples asserted that "he has absolutely struck out new lights, and not only adopted a breed of cattle and sheep, different from, and superior in many essential respects to most others, but established them in such a manner as to gain ground in every corner of Great Britain and Ireland, in consequence of their superior merit."[2] His only modern biographer began his book with the statement that "Robert Bakewell may be regarded as a man whose work assumed an importance which has not been exceeded by any agriculturalist before or since his time."[3] This assertion may smack of boosterism, but it is consistent with the more muted claims of relatively disinterested chroniclers. In the standard history of British livestock husbandry, Bakewell is the only individual to get his own chapter—or, indeed, to be mentioned by name in the table of contents.[4] His fireside chair, a capacious and solid wooden article, has been preserved as a trophy by the Royal Agricultural Society of England; it figured prominently under the rubric "The Agrarian Revolution: the Age of the Agricultural Improvers" in the large historical exhibition mounted to commemorate British Food and Farming Year in 1989.[5]

Like all legends, however, this one is open to reinterpretation, and on several levels. The heroic version of Bakewell's character and accomplishments is rooted in a heroic version of the agricultural revolution, which has itself come under increasingly energetic revision in recent years. That is to say, the complex of developments, embracing agricultural chemistry, crop selection and management, farm machinery and drainage techniques as well as animal husbandry, that was celebrated by eighteenth-century improvers as well as by subsequent antiquarians and historians as a great and discrete leap forward, is now frequently interpreted as part of a relatively protracted process with roots as far back as the medieval period.[6]

But the contestation of the Bakewell legend is not just an artifact of modern historiography. Bakewell was a controversial figure in his own time, inspiring detractors as well as admirers; the literature of contradiction and debunking has been as persistent, if not as voluminous, as the literature of adulation.

Possessing Mother Nature

Robert Bakewell's fireside chair.

Every component of his reputation has been repeatedly challenged, beginning with the most fundamental claim: that of his originality or priority. Arthur Young's widely echoed assertion that Bakewell's stockbreeding principles and techniques were "perfectly new" or at least "hitherto . . . totally neglected," coexisted with alternative stories about the creation of Bakewell's celebrated strains.[7] Sometimes these stories featured rival innovators. A shadowy predecessor named Joseph Allom was frequently named as the first improver of Leicester sheep, while a Mr. Webster of Canley was known to have done similar work with longhorn cattle, indeed, to have sold Bakewell several animals early in his career as an improver. Mr. Webster's old shepherd, it was rumored, also claimed that his master's ewes had helped found the Dishley flocks.[8] Other stories implicitly claimed that Bakewell had unfairly appropriated personal credit for accomplishments for which he was merely a synecdoche, by stressing the number of less celebrated contemporary farmers who had pursued analogous breeding strategies. Thus an eighteenth-century observer noted that the area around Dishley had "for many years abounded with intelligent and spirited breeders," and a nineteenth-century livestock historian claimed that the improvement of the longhorns "dates from the 1720s," the decade of Bakewell's birth.[9]

If Bakewell had not done it first, perhaps he had done it wrong. Critics claimed that while his principal technique, in-and-in breeding, or repeated crossing within a single lineage, might produce impressive results quickly, in the long run it led to delicate health and declining vigor; that is, that it "proved[d] as destructive to flocks, as marriages of near relations to the human kind."[10] Attacks on his system led to attacks on his judgment in less theoretical matters. Bakewell viewed sheep, cattle, and pigs as machines for producing meat, and so he selected them to maximize the expensive cuts and minimize the bones and innards, but skeptics alleged that the barrel shape he aimed at in breeding all species but the horse often induced an "excessive tendency to obesity," which "abates the procreative . . . powers."[11] And despite the striking changes he effected in the animals he chose for improvement, some of his strains proved to have surprisingly little staying power. Improved longhorns were fleshy, but their meat was unappealing, and as a side effect of improvement their milk yield and prolificacy decreased dramatically; after a brief heyday, they quickly disappeared from English pastures and slaughterhouses. Bakewell was no more successful than other stockbreeders in reforming the amorphous eighteenth-century pig. Even the renowned Dishley sheep might in retrospect come to seem more a triumph of public relations than of applied science.[12]

When they had finished with Bakewell's animals, critics might move on to his character. Although he was conventionally praised for the gentleness with which he treated his animals—it was said that even his largest bulls could be led by a child—and for the hearty hospitality of his bachelor table, the countertradition suggested that he drove a hard bargain, that he was meanspirited and possessive, and that his concern for profit led him to be secretive about information he should have shared. For example, when he considered his breeding rams to be past their reproductive prime, he sent them to the butcher; but to make sure that his colleagues and competitors would not attempt to reprieve them and thus gain free and unauthorized access to their carefully selected genetic material, he first infected them with "rot."[13] And he was as unwilling to part with this material in its abstract as in its concrete form; agricultural writers throughout the nineteenth century echoed contemporary complaints about "the mystery with which he . . . carried on every part of his business, and the various means which he employed to mislead the public."[14]

Damaging as such allegations might seem, however, they provoked no substantial riposte in more reverential discourse. On the contrary, they may have provided an occasion for the two traditions to converge. (Of course, these alternative commentaries also agreed that Bakewell was well worth talking about.) On the back of Bakewell's carefully preserved chair a nineteenth-century admirer carved the following inscription: "This chair was made under the direction of the Celebrated Robert Bakewell of Dishley out of a willow tree that grew on his farm. It was his favorite seat, and the back which thus records his Memory, served as a screen when seated by his fireside, calculating on the Profits, or devising some Improvement on his farm. Thousands of pounds have been known to exchange hands in the same. . . ."[15] As a eulogy this may leave something to be desired, but it nevertheless suggests the source of Bakewell's enduring renown and notoriety. Whether or not his selective breeding techniques reshaped his longhorn cattle and Leicester sheep as much or as well as he and his adherents have claimed, there is no question that, on the conceptual level, they helped to restructure the whole enterprise of animal breeding.

Bakewell assumed that the qualities that defined animal excellence were inherited and that careful selection and pairing of parents would ensure that their own desirable attributes were replicated or enhanced in their offspring. That is, he assumed that it was possible for the improver to redraw the conventional boundary between the sphere of nature and the sphere of agriculture. These assumptions were not necessarily inconsistent with received stockbreeding wisdom; indeed, there was no contemporary consensus about what could be inherited and how.[16] But the prevalent practices of mid-eighteenth-century husbandry were based on other assumptions, particularly the predominance of such environmental factors as climate and diet in determining the qualities of adult animals. Even at the end of the eighteenth century, the distinguished members of the Board of Agriculture devoted a great deal of attention to the effect of different grazing regimens on sheep and cattle breeds.[17] The breeders of a few kinds of luxury and sporting animals—mainly racehorses, and to a lesser extent foxhounds and greyhounds—selected and paired their stock along the lines later adopted by Bakewell, but most breeders of farm animals were less concerned with the heritable merits of individual animals.[18] What they wanted from their herds and flocks, in addition to meat or milk or wool or tallow, was a healthy and numerous progeny, not particularly distinguishable from their parents or from one another. As far as reproduction was concerned, parent animals were also essentially interchangeable. Breeding males were selected on the basis of availability and willingness to procreate; breeding females, by and large, were not selected at all. The result was a livestock population consisting mostly, at least in Bakewell's terms and those of subsequent improvers, of rather nondescript regional strains.[19]

Bakewell's system replaced fungibility with a high degree of differentiation. He assessed the reproductive potential of individual animals in subtle qualitative terms rather than in terms of simple addition. It became possible to ask not merely how many but also how good and how reliable. The answers to these new questions were not, however, merely qualitative; augmented quality meant augmented value, which was routinely measured in cash. Perhaps the most telling evidence of the desirability of Bakewell's stock and the prestige of his method was the amount other breeders were willing to pay for the services of his male animals, especially his rams.[20] When Bakewell began renting out rams, or tups, in 1760, he charged 17s. 6d. per animal per season (this fee included as many ewes as the ram could be persuaded to serve), which can be roughly compared with bulling fees of between 6d. and 1s. per cow, which were standard throughout much of the eighteenth century; and with the stud fee of a guinea per mare advertised in 1755 by the owner of several undistinguished stallions.[21] By 1784 Bakewell was asking £100 per season for his best rams, and five years later equivalent services commanded £300–£400 per season. (His rule for bulls, once he had hit his marketing stride, was to ask half the animal's value for a season's hire.)[22]

The breathtaking steepness of this trajectory—a fourhundredfold increase within thirty years—suggests that it recorded a change in kind rather than (or as well as) a change in degree. The escalation in fees was too rapid to reflect simply the enhanced quality of the New Leicesters, striking though that was in the view of many; it also represented the entry of a whole new source of value into the price calculation. Bakewell claimed that when he sold one of his carefully bred animals, or, as in the case of stud fees, when he sold the procreative powers of one of these animals, he was selling something much more specific, more predictable, and more efficacious than mere reproduction. In effect, he was selling a template for the continued production of animals of a special type: that is, the distinction of his rams consisted not only in their constellation of personal virtues, but also in their ability to pass this constellation down their family tree. They were the result of a minor act of creation, and if they were skillfully managed, they could be the agents or catalysts of additional similar acts. Thus it was possible for a disciple like George Culley, of Durham, to transform his own flocks by hiring a ram from Bakewell each year. He began this practice in the 1760s, when Bakewell was just starting to make his reputation, and continued it even after other northern breeders had begun paying liberal fees to hire Culley rams.[23] And this redefined relation between an individual animal and its lineage was newly reciprocal as well as newly powerful. At least the best breeding animals were expected to enhance as well as merely to express and pass on the reproductive potential of their strain, a point that was emphasized by the tendency of some breeders to name lineages (often referred to as "families" or "tribes") after particularly influential or exemplary ancestors.[24]

So complete was the conceptual transformation wrought by this redefinition of an animal's worth that at a remove of two centuries it may be difficult to recover its novelty. We now routinely take the kind of genetic property created by Bakewell and his adherents into account when pricing animals; indeed, often it is the primary, if not the only, source of value—the reason, for example, that neutered animals are not permitted to compete for most show prizes. Nor does this policy reflect a recent consensus. If the possibility of theft may be taken to indicate the presence of property, then the concept of genetic property was already firmly entrenched a hundred years ago, when the Kennel Club was formed to regulate dog showing and breeding. At least in its early years, only a few violations were considered heinous enough to be censured by the club's governing committee, but stolen kisses were definitely among them. Two cases of genetic larceny immediately occupied the committee's attention: One was committed by a breeder who brought his bitch in heat to a dog show (already an infraction, since animals in that condition were officially barred) and managed to position her within the easy reach of a male champion. The other was committed by a breeder whose attempts to purchase the services of a particular dog had been spurned by that animal's owner. By dint of spying, plotting, and the corruption of railway employees, he arranged for his bitch and the dog in question to spend some unsupervised moments together in a baggage compartment.[25] Although such behavior had been recognized as theft since at least the end of the eighteenth century, the Kennel Club disciplinary hearings may have represented a hardening sense of the seriousness of the offense. When Thomas Booth, one of the best-known early improvers of shorthorn cattle, was similarly imposed upon—by a tenant who put an attractive cow into the field next to one of his most distinguished bulls, thus depriving Booth not only of a stud fee but also of the cost of repairing the fence—it was remarked more as an instance of ludicrous delinquency than as one of serious moral turpitude.[26]

But even this rather lighthearted acquiescence in the possibility of owning the design of an animal represented a significant shift in stockbreeders' attitudes, a shift in which the miscreant participated, since Booth, like many progressive landlords, provided the services of the more ordinary bulls to his tenants gratis. Forty years earlier, Bakewell had met with a great deal of resistance when he instituted his policy of hiring out his best rams for the season. In the words of David Low, the professor of agriculture at the University of Edinburgh during the first part of the nineteenth century, "the plan was ridiculed and opposed in every way, and it was not until after the lapse of nearly a quarter of a century that he succeeded in establishing it as a regular system."[27] This resistance was probably only partly a kneejerk conservative reaction to a perceived commercial innovation. After all, although Bakewell's name was strongly associated with seasonal hiring—for example, in 1790 the engraver and naturalist Thomas Bewick illustrated "the modern practice of letting out Rams for hire by the season; which from very small beginnings, has already risen to an astonishing height" exclusively with reference to Bakewell—he was, in this case as in others, only the publicizer and standardizer, not the innovator.[28] Other breeders had rented out rams without provoking hostile commentary, and, in any case, the practice bore an obvious relation to the stud fees traditionally charged for individual services.

Nor can the criticism of seasonal hiring be blamed on Bakewell's prices, since these did not rise high enough to provoke complaints until the 1780s, when resistance to his marketing techniques had subsided. Instead, since Bakewell linked the practice of hiring with the special virtues he claimed for his stock and with his reluctance to alienate any of his best animals permanently, it is more likely that the strong initial hostility to renting rams by the season reflected a suspicion of the new notion of property embodied in Bakewell's system. Potential customers may have worried that he was selling something that should not or, to put it more skeptically, could not be sold.

His marketing strategy seemed designed to reify the somewhat intangible genetic property embodied in his animals. He formalized this strategy (and included within it his closest colleagues and rivals) in 1783, when he founded the Dishley Society. The high-end breeders of New Leicester stock who composed its membership agreed to comply with an elaborate and stringent set of restrictions on their activities, including the number of rams they could let per season, when they could let them, and to whom they could let them. It was, for example, forbidden to let rams to anyone who let or sold rams at fairs or markets. (Other prohibitions included letting rams on Sundays and leaving society meetings without asking permission.) Prices were fixed too: at least 10 guineas per ram per season to ordinary breeders, at least 40 guineas to breeders who themselves let rams, and at least 50 guineas for any ram let by Bakewell himself to anyone living within one hundred miles of Dishley. In return for compliance (which was enforced by fines of up to 200 guineas) and for dues of about 100 guineas, members received privileged access to each other's and especially to Bakewell's stock, as well as the benefits of belonging to a cartel.[29]

The Dishley Society may have been the most exclusive marketplace for highly engineered genetic material in the 1780s, but it was not the only one. Persuaded either by the sight of the improved animals, which were available

Illustration by George Garrard.

for inspection at the farms of breeders and sometimes at fairs or semipublic exhibitions like the sheepshearings sponsored by elite agriculturists, or by their promotion in the agricultural press, stockbreeders gradually accepted the notion of genetic property implicit in Bakewell's practices. Demand for this commodity consequently burgeoned, and, predictably enough, so did the supply. (This increase in supply and demand was not restricted to New Leicesters, but extended to other kinds of sheep, and to improved strains of other domesticated species, especially cattle.)[30] Once breeders proliferated, quality control became an issue. Bakewell himself had refused to give any guarantee of the quality of his stock except his own name. Not only was he loathe to reveal the parentage of particular animals, but, as even his admirers have lamented for the past two centuries, he systematically concealed the breeders and the regions that had supplied his foundation stock.[31] But no one else could get away with this kind of metonymic assurance. Lesser men had to offer harder evidence.

The template that constituted the chief value of an expensive animal was a rather tenuous article, not immediately apparent to the prospective purchaser or renter. In the case of uncomplicated animals that had not been subject to Bakewellian reconstruction, a shrewd eye and hand could assess the quality of an adult with some confidence, and make a fair guess as to the probable development of a juvenile. But such tools failed when the issue was reproductive power rather than personal charm. The two most important components of an animal's genetic endowment—the best indication of its likelihood of passing on desirable qualities to its progeny—were both functions of its lineage: purity of descent, meaning a heritage that included a preponderance of forebears with the same qualities; and prepotency, meaning a heritage sufficiently concentrated and powerful to dominate the heritage of potential mates. As stockbreeding manuals repeatedly advised, when the production of distinguished offspring was the main point, these hidden qualities could be more important than the manifest ones. The annals of husbandry included many rams, bulls, and stallions who had won no prizes themselves but were celebrated "getters." As one mid-nineteenth-century cattleman put it, "We ought always to prefer a bull of high pedigree, with fair symmetry and quality, to another bull, though much superior in appearance, but of questionable pedigree."[32]

Bakewell's contemporaries would have appreciated the wisdom of this advice, although they might not have put it in the same terms, since the *Oxford*

"The Leicestershire Improved Breed," from
Thomas Bewick, *General History of Quadrupeds*, 1824.

English Dictionary does not report the word *pedigree* being applied to domestic animal lineages until the first part of the nineteenth century. And difficulties of implementation would have overshadowed those of expression. The acceptance of paper pedigrees as reliable evidence of the existence, legitimacy, and value of genetic property depended on elaborate, plausible, and sustained record-keeping. Such records were hard to come by in mid-eighteenth-century Britain. Indeed, before the new and more nuanced sense of an animal's individuality implied by Bakewell's practices, they were almost inconceivable. As long as domestic animals were understood as interchangeable members of flocks and herds, they were rarely named; their reproductive history was apt to be obscure, and their genealogy impossible to trace.[33] Even the renowned imported stallions who laid the foundation for modern thoroughbred horses in the late seventeenth and early eighteenth centuries were, somewhat paradoxically, semianonymous, referred to only by their English ownership and their supposed country of origin: the Darley Arabian, the Godolphin Barb, the Byerley Turk. One of the most celebrated animals of the 1790s, a massive specimen of the shorthorn breed of cattle, toured simply as "the Durham ox."

Speculative reconstructions of the descent of Bakewell's own animals include such shadowy creatures as "a Westmoreland bull" and "a Canley cow," said to be the parents of the distinguished longhorn bull Twopenny, who figured prominently (and also frequently, due to Bakewell's predilection for inbreeding) in many subsequent pedigrees.[34] Although he did not publicize his animals' family trees, Bakewell needed to keep track of them for his own continuing experimental purposes. And so his breeding stock, like that of many of his contemporaries, emerged into the light of nomenclature. By the end of the eighteenth century, any well-bred animal was likely to come supplied not only with its own name, but with those of its parents. Often these names injected a little poetry into the ordinarily prosaic practice of animal husbandry. For example, their portraits proclaimed that the shorthorn Comet was the son of Favourite and Young Phoenix, and that the longhorn Garrick was the son of Shakespear and Broken Horn Beauty. None of these animals had been named by Bakewell, however, whose taste in nomenclature, as in other things, tended toward the functional: many of his sheep were identified merely by letters of the alphabet or by butchers' terms that suggested their strongest marketing points, for example, Shoulders, Bosom, Carcass, and Campbells (hocks).[35]

Thus the value of the most distinguished animals was ratified in language after it had been created in the breeding pen; in a sense, it did not exist on the hoof until it had been inscribed in the record books. But although written evidence might be a necessary guarantee of genetic property, it was not sufficient. The breeders' private records were only too liable to manipulation and falsification, to the creation of genetic property in a different and more radical sense, with no help from the breeding pen. Occasionally the pedigrees of superstar animals such as Comet and Garrick would be published in agricultural or sporting periodicals, which at least put a stop to their further elaboration. This practice was generalized in the breed books, really genealogical catalogues, that began to appear at the end of the eighteenth century. The preface to the first of them, the *General Stud Book* for racehorses or thoroughbreds, published in 1791, defined its purpose as "to correct the . . . encreasing evil of false and inaccurate Pedigrees"; the preface to the *General Short-Horned Herd Book* of 1822, the first such work devoted to farm livestock, noted that "it must be both the interest and the wish of every breeder to be enabled to breed with the greatest possible accuracy as to pedigrees."[36]

More was at stake in this interest and wish than merely forestalling fraud. Stud books proliferated during the nineteenth century, even for breeds in which misrepresentation could hardly be considered an issue; that is, breeds that had been little improved, and for which, therefore, the problem facing cataloguers was an absence of information rather than a plethora. For example, the secretary of the British Berkshire Pig Society assured readers of its first *Herd Book* that all animals listed within it were "of undoubted purity," even though many of them had been included on the basis of "exceedingly short pedigrees" or "only a note"; the compiler of the first *Ayrshire Herd Book* complained that he had been impeded in his labors "as hitherto names have rarely been given to animals."[37] The editors of the first *British Goat Society Herd Book* may have faced the most difficult task of all, since "the propagation of these animals has not been conducted with any regard to purity of breed" and "very few breeders . . . have kept any record of Pedigree."[38] Herculean though they may have been, however, these labors were rarely thankless. Indeed they were predictably if indirectly rewarded in cash. The editors of the second volume of the *Sussex Herd Book* complacently noted that in the six years that had passed since the appearance of the first volume in 1879, the prediction there expressed "that Sussex Cattle would come more and more into favour, had been fully borne out, and we can point with satisfaction to the greatly increased number of Animals."[39] Increased "favour" meant larger demand, and consequently higher profits for breeders.

The enhanced value that normally followed the appearance of a breed book may have been partly the result of the attendant publicity's leading to a wider appreciation of the merits of a particular kind of stock. But breed books created distinction as well as merely cataloguing it. As the establishment of individual pedigrees defined and reified the genetic endowment and capacity of individuals, group genealogies analogously defined and reified the template associated with a given breed. And this definition could work both ways. That is, if the existence of a numerous and popular breed required the compilation of a breed book, so the compilation of a breed book could imply the existence of a breed. Thus, members of the Galloway Cattle Society felt that "it was desirable to have a separate Herd Book" for their animals, even though they had probably "sprung from the same source" as the Polled Angus and Aberdeens; with similar determination, the editor of the first *Norfolk and Suffolk Red Polled Herd Book* made a point of denying the assertion that "this is but a branch of the Galloway breed naturalized here."[40] The *Oxford Down Flock Book* was both evidence and agent of its society's "special pride and boast": to have created an independent breed from the combination of two others.[41]

The literary construction of breeds could have a physical dimension.

"Short-Horned Bull," from David Low, *Breeds of the Domesticated Animals*, 1842.

Especially for breeds that had been only loosely defined—breeds that, like many local strains of cattle, sheep, and pigs, consisted of a regional label and little else—the publication of a catalogue might encourage breeders to mate their animals within the prescribed pool and to select animals possessing the characteristics identified as most desirable. The British Goat Society offered an extreme example of this possibility when it confessed, in its first *Herd Book,* that it could not aspire to "preserve a record of pure-bred Stock," only to guide "breeders desirous of introducing fresh blood into their herds."[42] But most creation by breed book occurred on the conceptual level. Like the process of individual naming and genealogy that they repeated and multiplied, these catalogues concretized a rather abstract component of the value of the animals listed within them. As an animal's lineage enhanced and guaranteed the promise of its personal qualities, so its membership in a breed underwrote the claims of its lineage.

And although their ostensible function was straightforwardly descriptive, breed books helped to redefine the category they illustrated. After all, despite the complaints of some elitists that they were not discriminating enough, they did not list every animal proposed for inclusion.[43] Decisions to admit inevitably both reflected and reinforced the compilers' idea, not only of the nature of thoroughbred horses or shorthorn cattle, but of breed itself. By the middle of the eighteenth century, this term had become unstable. As the *Oxford English Dictionary* suggests, the use of the word *breed* to refer to a related stock of animals, along with the more or less interchangeable constellation of *race, strain,* and *variety* (the latter often carrying a scientific connotation, but no real difference in denotation), had been common at least since the late sixteenth or early seventeenth century. Originally what defined a breed seems to have been shared provenance and shared function, as well as some degree of physical resemblance. As late as the eighteenth century the term was often used in that sense by enlightened agriculturalists who wished to disparage unimproved animals. Thus, in 1794, a surveyor commissioned by the Board of Agriculture referred to the "original black breed" of Carmarthenshire cattle as "ill shaped and unprofitable to the pail," and to the local "breed of swine" as "a narrow, short, prick-eared kind."[44]

When improved animals were in question, however, animals of the same breed might have both more and less in common. Geographical contiguity was only accidental and so was close familial connection, except in breeds based on intense inbreeding, such as the New Leicester sheep, or breeds in which a few ancestors had been particularly appealing, such as the shorthorns, all of which were descended in one way or another from Comet.[45] What was essential, however, was the template. As one improving farmer put it, certain "varieties have been usually distinguished among farmers by the appellation of different *breeds;* as they have supposed that their distinguishing qualities are, at least, in a certain degree, transmissible to their descendants."[46] Such definitions bestowed upon breeds a newly enhanced degree both of permanence and of content.

As a result, the category represented by *breed* increased in status. And this glorification was not apparent only within the discourse of elite animal husbandry, where cynics might ascribe it to narcissism or financial interest. The consensus of enlightened agriculturalists was echoed by another discourse, one that was further removed from the barn and the butcher and that lent the authority of science. Domestic animals belonged to the animal kingdom as well as to the enterprise of agriculture, and they figured prominently in the catalogues and the schemata of quadrupeds voluminously produced by eighteenth-century naturalists. Indeed, because they were easy to observe and economically important, they often occupied far more than their share of space.[47] They were favored not only in terms of pages allotted, but also in terms of categorical analysis.

Although eighteenth-century taxonomies differed from modern ones in many important ways, involving both form and content, they similarly consisted of a hierarchical—that is to say, increasingly inclusive—set of categories. But not every animal group, and especially not every group of domestic animals, found a place in this structure; to be included at all conveyed a measure of distinction. The smallest category conceived to represent significant differentiation was the species. Designation as a species corroborated and validated a group's separateness. In addition, since the form of hierarchy inevitably suggests increasing prestige, species possessed a kind of status that mere races, strains, or varieties lacked. Although drawing the boundary between any particular species, as well as defining *species* in general, presented, then as now, very knotty problems, pragmatic differentiations were ordinarily based on physical similarity and the production of fertile offspring.

A rigorous application of those criteria should not have privileged breeds of domestic animals. Different breeds of cattle or pigs or dogs resembled one another physically and reproduced with ease and efficiency; and in any case increased variability within species was widely recognized as one of the most frequent consequences of domestication. Nevertheless, naturalists tended to accord breeds of domestic animals the same taxonomic status as species of wild animals. Sometimes this was done tacitly. For example, in his popular *General History of Quadrupeds* (1790), Thomas Bewick presented "The Arabian Horse," "The Race-Horse," "The Hunter," "The Black Horse," and "The Common Cart House" in separate entries analogous to those devoted to "The Ass" and "The Zebra."[48] Following Linnaeus, George Shaw tagged many dog breeds with Latinate binomials that at least sounded like the names of species; for example, the hound was *Canis sagax,* the shepherd's dog was *Canis domesticus,* and the Pomeranian was *Canis pomeranus.*[49] And if the same claim could not quite be made for cat breeds, another disciple of Linnaeus presented minor feline variations as subspecies; thus, the angora cat was *Felix catus angorensis* and the tortoiseshell was *Felis catus hispanicus.*[50] The *Naturalist's Pocket Magazine* argued that treating certain kinds of domestic sheep "as varieties, does not . . . sufficiently discriminate between them. . . . To us it appears, that there is, probably, even a specific distinction."[51] Thomas Pennant, a distinguished zoologist, upped the ante still further. He referred to what were known as the wild cattle of Chillingham, an unruly strain of white animals preserved in the parks of several great houses, mostly for decoration or sport rather than for the dairy or the slaughterhouse, as *Bisontes scotici,* putting them into a separate genus or subgenus from the *Bos taurus* that grazed on ordinary British pastures.[52]

Given this ball, some agriculturalists quickly ran with it. Claims for the taxonomic reification of breeds might be made obliquely, as when Bakewell's disciple George Culley suggested with regard to domestic goats that "the different species . . . might be greatly improved, by the simple rule of selecting the best males and best females."[53] Or they might be made circumspectly, as when John Wilkinson asserted that "The distinction between some [breeds] . . . has scarcely been less than the distinction between that variety and the whole species. The longer . . . these perfections have been continued, the more stability will they have acquired and the more will they partake of nature."[54] But in his comprehensive manual of farm livestock, John Lawrence distinguished directly between the genus, which included "original and distinct kinds . . . as the genus of neat cattle, of swine, of sheep," and the species, which included "the most remarkable divisions of genus: as in the horse genus, the racer and the cart horse; in neat cattle, the bison and the common European species, also the long and shorthorned: in sheep, the coarse and long, the fine and short woolled: in swine, the lop and prick eared."[55]

Thus the distinctions that separated breeds of domestic animals were conceptualized in terms that made them competitive with, and sometimes identical to, the impenetrable barriers that were generally held to divide wild species. The accomplishments of agricultural improvers might thus rival as well as shadow those previously wrought by nature. This implied that something very real and valuable indeed had been created, since, as one English translator of Buffon put it, "the most constant and invariable thing in Nature is the image or model allotted to each particular species."[56] And more was implied in this claim than the creation of new genetic property for stockbreeders to buy and sell, important though this was. The implicit removal of domestic animals from the natural realm into the realm of technology recorded a shift in the relation between people and their natural environment—an expansion of the territory in which people need not fear to tread.

All this may go some way toward explaining the long historical shadow cast by Robert Bakewell, even if he did not accomplish any of the things on which his legendary stature is said to depend. Whether or not he really improved the longhorns and New Leicesters, or whether anyone did, or whether they were any good after they had been improved—whether he introduced new commercial methods or merely borrowed those used by others with less effective publicists—he was both agent and symbol of a more profound, if less tangible, shift. The market in animal templates that emerged in consequence of the wide appreciation of his labors, if not of those labors themselves, depended on a reconceptualization of the kind of property that an animal constituted, which itself implied an enlargement of the appropriate sphere of cultural activity. Such reconceptualizations of the relationship between culture and nature, between people and God, tended to stir things up two hundred years ago, and they have not lost that power. And the prominence of the profit motive has usually intensified the uneasiness provoked by such boundary shifts—when humans transcend their biblically allotted role as names to become creators. Thus it should be no surprise that many of the issues raised by Bakewell's work, and much of the energy in the debate it provoked, have resurfaced only superficially transformed in recent and continuing discussions of genetic engineering, recombinant DNA, and the patenting of engineered organisms.

Notes

I am grateful to Juliet Clutton-Brock for her comments on an earlier version of this essay.

1. Pawson, *Robert Bakewell*, 65–66.
2. Culley, *Observations on Live Stock* (1786), 26.
3. Pawson, *Robert Bakewell*, xiii.
4. Trow-Smith, *History of British Livestock Husbandry*.
5. Spargo, *This Land is Our Land*, 41. The year 1989 was both the 150th anniversary of the first Royal Agricultural Society show and the centenary of the creation of the Ministry of Agriculture, Fisheries and Food.
6. Nicholas Russell's *Like Engend'ring Like* offers the most elaborate revision of the traditional history of stockbreeding. See also John Walton, "Pedigree and the National Cattle Herd"; and Edwards, *Horse Trade*.
7. Young, *Farmer's Tour Through the East of England*, 1:110, quoted in Bajema, *Artificial Selection*, 29.
8. Ryder, *Sheep and Man*, 486; Russell, *Like Engend'ring Like*, 146; Thomas Barnet to Colonel Robert Fulke Greville, 30 December 1789, in Banks, *Sheep and Wool Correspondence*, 179. See also Russell, "Who improved the eighteenth-century longhorn cow?" 19–40.
9. W. Marshall, *Rural Economy of the Midland Counties* (1790), quoted in Chambers and Mingay, *Agricultural Revolution*, 67; John Coleman and Gilbert Murray, "The Longhorns," in Coleman, *Cattle of Great Britain*, 83.

10. Blacklock, *Treatise on Sheep*, 102. This reservation, it should be emphasized, was technical rather than moral. In-and-in breeding drew occasional criticism on the ground of incest, but such commentary was not taken very seriously by Bakewell and his disciples. Ritvo, *Animal Estate*, 67.

11. John Lawrence, *General Treatise* (1805), 387.

12. Low, *Breeds of the Domestic Animals*, 1:48; Hall and Clutton-Brock, *Two Hundred Years of British Farm Livestock*, 203–4; Professor Sheldon, "Sheep," in John F. L. S. Walton, *Best Breeds of British Stock*, 110; Russell, *Like Engend'ring Like*, 215.

13. "Rot" was probably liver fluke rather than foot rot, as has sometimes been alleged. Pawson, *Robert Bakewell*, 83–84.

14. Sebright, *Art of Improving the Breeds of Domestic Animals*, 9.

15. Spargo, *This Land is Our Land*, 41.

16. For scientific views, see Farley, *Gametes and Spores*, chap. 1.

17. This is a recurrent theme in the Board of Agriculture minute book, 27 November 1798–18 March 1805.

18. Ritvo, "Pride and Pedigree," 230–33.

19. Ritvo, *Animal Estate*, 64–66.

20. For Bakewell's bull-hiring, see James Wilson, *Evolution of British Cattle*, 120–21.

21. Low, *Breeds of Domestic Animals*, 2:67; Russell, *Like Engend'ring Like*, 152; newspaper clipping in John Johnson Collection of Printed Ephemera.

22. Low, *Breeds of Domestic Animals*, 2:67; Bates, *Thomas Bates and the Kirklevington Shorthorns*, 40.

23. Russell, *Like Engend'ring Like*, 210–11; Sinclair, *Code of Agriculture*, 95.

24. One of the best-known examples was the Duchess line, established by Thomas Bates early in the nineteenth century. Bates and Bell, *History of Improved Short-horn or Durham Cattle*, 209; Hall and Clutton-Brock, *Two Hundred Years of British Farm Livestock*, 50.

25. Kennel Club minute book. December 1, 1874 to April 21, 1884.

26. William Carr, *The History of the Rise and Progress of Killerby, Studley and Warlaby Herds of Short-horns* (1867), quoted in Lewis Falley Allen, *History of the Short-horn*, 101–2.

27. Low, *Breeds of the Domestic Animals*, 2:67.

28. Bewick, *General History of Quadrupeds* (1822), 64; Russell, *Like Engend'ring Like*, 208–9.

29. Pawson, *Robert Bakewell*, 72–78.

30. See, for example, John R. Walton, "Diffusion of Improved Sheep Breeds"; and Clutton-Brock, "British Cattle in the Eighteenth Century."

31. Many subsequent commentators have speculated about this mystery. See, for example, Hall and Clutton-Brock, *Two Hundred Years of British Farm Livestock*, 151; Trow-Smith, *History of British Livestock Husbandry*, 60–62; Russell, *Like Engend'ring Like*, 208–10; and Ryder, *Sheep and Man*, 486.

32. M'Combie, *Cattle and Cattle-Breeders,* 152–53.

33. Keith Thomas attributes the practice of naming pet animals to a parallel process of individuation. *Man and the Natural World,* 113–14.

34. James Wilson, *Evolution of British Cattle,* 119.

35. Spargo, *This Land is Our Land,* 42–43, 88; Russell, *Like Engend'ring Like,* 213.

36. *General Stud Book,* iii; Coates, *General Short-Horned Herd-Book,* vii.

37. *British Berkshire Herd Book,* vii; Vernon, *Ayrshire Herd Book,* v.

38. *British Goat Society Herd Book,* 1.

39. *Sussex Herd Book,* preface, n.p.

40. *Galloway Herd Book,* preface, n.p; *Norfolk and Suffolk Red Polled Herd Book,* 9.

41. *Oxford Down Flock Book,* 8.

42. *British Goat Society Herd Book,* 2.

43. One such elitist was Thomas Bates, who declined to list his shorthorns in the *Herd Book* after his friend and its founding editor, George Coates, died, on the ground that he did not want his highly bred animals flanked by mongrels. Ritvo, *Animal Estate,* 61–62.

44. Hassall, *General View of the Agriculture,* 35, 37.

45. James Wilson, *Evolution of British Cattle,* 124.

46. James Anderson, *Essays Relating to Agriculture and Rural Affairs,* 138–39. This definition and others along the same lines did not settle the matter completely. Disagreement and confusion about the nature of breeds has persisted to the present. For the problem and the solution, see Clutton-Brock, "Definition of a Breed."

47. Ritvo, *Animal Estate,* 18.

48. Bewick, *General History of Quadrupeds,* 1–23.

49. George Shaw, *General Zoology,* vol. 1, pt. 2, 277–80.

50. Kerr, *Animal Kingdom or Zoological System,* 154.

51. *Naturalist's Pocket Magazine,* n.p.

52. Pennant, *History of Quadrupeds,* 1:16–17.

53. Culley, *Observations on Live Stock* (1786), 220.

54. John Wilkinson, *Remarks on the Improvement of Cattle,,* 4–5.

55. John Lawrence, *General Treatise* (1805), 1.

56. Leclerc, *Barr's Buffon,* 48.

Our Animal Cousins

It is a historical commonplace—it might even be called a factoid—that the publication of *On the Origin of Species* in 1859 marked a kind of watershed in the understanding of the relationship between humans and (as we now say it) other animals, at least in the Anglophone world, and possibly in the West more generally. In other words, it is assumed that a definite gap, one that had existed previously, was bridged at that point. But this gap may not have been so definite, or even so preexisting, as this widely accepted formulation suggests. Even in the Victorian or pre-Victorian context, the shift from the human to the nonhuman animal frequently required no very great leap. On both the physical and the rhetorical level the relationships between people and animals had long seemed indeterminate and fluid; to put it another way, determining the point where unlikeness became more significant than likeness had always been problematic. At the most basic level, there was no consensus about the content of the dichotomy that distinguished people from the rest of the animate creation. Uncertainty routinely ran in both directions. Some members of what we recognize today as the human species could be relegated to the category "animal," while some members of other species could be included within the human circle.

Both political and scientific discussions of the population of Great Britain, for example, frequently suggested that the Irish might be less fully human than the Scots, and the Celts less fully so than the Saxons. Nineteenth-century iconography was rich in comparisons between what was portrayed as the ste-

---

"Our Animal Cousins" originally appeared in *Differences* 15, no. 1 (1994): 46–68.

reotypical Irish physiognomy and that of the great apes. And non-Europeans were still likelier to be the subjects of such taxonomic discrimination, as were, within any given society, women and children. On the other hand, Victorian commentators often compared well-bred horses and dogs favorably with what they regarded as inferior human types, on grounds of intellect, as well as of disposition and personal appearance. And, then as now, if people were asked directly whether they felt themselves to be closely related to (other) animals, they were liable to give a different answer (no) than would be concluded on the basis of observing their behavior or eavesdropping on their conversations (yes).

This profound inconsistency helped to structure the most powerful and self-conscious figurative expression of relationship: zoological classification. From its Enlightenment beginnings, most formal taxonomy recognized not only the general correspondence between people and what were then known as quadrupeds, but also the more particular similarities that human beings shared with apes and monkeys. (It was the nonfunctional physical details that often proved most compelling: the shape of the external ear, for example, or the flatness of fingernails and toenails.) Thus, in 1699 the anatomist Edward Tyson published a treatise entitled *Orang-outang, sive Homo Sylvestris. Or the Anatomy of a Pygmie compared with that of a Monkey, an Ape* [by which he meant a baboon] *and a Man*. Tyson stated in his preface that his purpose was to "observe *Nature's Gradation* in the Formation of *Animal* Bodies, and the Transitions made from one to another," thus implicitly including humanity in the animal series.[1] Not only did Tyson present people as anatomically continuous with animals, but his choice of terminology further implied that the categories "human" and "orangutan" might not be completely distinct. Both of his synonyms for the orangutan (by which he meant what is now known as a chimpanzee) mentioned in the title conflated it with people: the translation of *Homo Sylvestris* is "wild man of the woods," and, conversely, the humanity of the quasi-mythical pygmies had long been the subject of European speculation. Even at the end of the eighteenth century, naturalists could claim that the "race of men of diminutive stature" or the "supposed nation of pygmies" described by the ancients, was "nothing more than a species of apes . . . that resemble us but very imperfectly."[2]

The celebrated eighteenth-century systematizer Carolus Linnaeus also located people firmly within the animal kingdom, constructing the primate order to accommodate humans, apes, monkeys, prosimians, and bats. In what

Wild orangutans drawn to look human,
from Edward Donovan, *Naturalist's Repository*, 1822–24.

has become the definitive (1758) edition of his *Systema Naturae*, he included two species within the genus *Homo*. One was *Homo sapiens*, subdivided into (mostly) geographical subspecies or races, such as *H. Sapiens Americanus* and *H. Sapiens Europaeus*, and the other was *Homo Troglodytes*, which was also known, Linnaeus pointed out, as *Homo sylvestris Orang Outang*.[3] Thus Linnaeus grouped Tyson's chimpanzee with humanity, rather than including it in the crowded genus *Simia* with the monkeys and the other apes.

And this taxonomic metaphor of connection was not necessarily confined to the realm of abstraction; through passion or sentiment it might be embodied in living flesh. The birth of hybrid infants has conventionally (although always problematically) been taken to indicate identity of species on the part of the parents. Crossbreeding has always tended to engage both vulgar and learned curiosity. Most reported hybrids involved nonhuman animals—including cattle and bison, dogs and wolves, horses and zebras—even such unlikely pairs as sheep and raccoons. Among the most celebrated zoological attractions of the first half of the nineteenth century—ballyhooed as "Unparalleled Attraction! Prodigies of Nature!" on one provincial tour—were the multiple litters that resulted from an unusual friendship between a male lion and a female tiger in Thomas Atkins's traveling menagerie.[4] At midcentury the "leporides"—a large family of alleged rabbit-hare hybrids bred in France—riveted the attention of British journalists and scientists.[5]

Humans too could be the objects or the originators of passions that tran-

"Plate LX. Rodentia," from Hugh Craig, *The Animal Kingdom*, 1903. Leporides were claimed to be rabbit-hare hybrids.

scended or violated the ostensible species barrier, although accounts of such amorous episodes tended to be carefully distanced by skepticism or censure. And there were other ways of positing similarly concrete connections between people and the nonhuman animals most nearly allied to them by anatomy. Well into the nineteenth century, physicians explained many kinds of birth defects as the unfortunate consequences of what was termed maternal imagination or impression—that is, of the expectant mother's fascination with an external object that had somehow influenced the development of her unborn child. Where the object was animate, it could occasion a kind of mental hybridization—a child whose parentage involved more than one species. Thus in 1867 the *Lancet* attributed the dense fur that covered an unfortunate girl's back to the fact that her mother had been frightened by an organ grinder's monkey.

In addition, commitments that were explicitly or essentially theological made many naturalists reluctant to locate their own species within the system of animal connections, whether it was figured in the Enlightenment mode as a chain, or in the Victorian mode as a tree. Some dissenters simply proposed their own countertaxonomies, which implicitly posited a much wider separation. Thus early in the nineteenth century the anatomist William Lawrence suggested that "the principles must be incorrect, which lead to such an approximation" (that is, between humans, apes, and monkeys in the primate order); he argued instead that "the peculiar characteristics of man appear to me so very strong, that I not only deem him a distinct species, but also . . . a separate order."[6] Naturalists who recognized this exclusively human order normally designated it as "Bimana," which stressed the erect posture and purpose-built feet characteristic of people, in contrast with the four-handed apes and monkeys who were segregated in the order "Quadrumana."[7]

At the same time that evolutionary theory suggested a more concrete and ineluctable connection, it provoked still more forceful resistance. The geologist Adam Sedgwick, who had been one of Darwin's early scientific mentors at Cambridge, anonymously expressed his "deep aversion to the theory" of evolution by natural selection on taxonomic grounds. He asserted that "we cannot speculate on man's position in the actual world of nature, on his destinies, or on *his origin,* while we keep his highest faculties out of our sight. Strip him of these faculties, and he becomes entirely bestial; he may well be (under such a false and narrow view) nothing better than the natural progeny of a beast, which has to live, to beget its likeness, and then die for ever."[8] As

Darwin himself sadly noted at the end of *The Descent of Man,* written a decade after the appearance of *The Origin,* "The main conclusion arrived at in this work, namely that man is descended from some lowly-organised form, will, I regret to think, be highly distasteful to many persons."⁹

And the scientific resistance to classifying people as primates was also echoed in the visual tradition. One way of denying the human-ape connection was to posit an alternative alliance. If nonprimate animals resembled humans more closely than apes, then they would necessarily displace apes from their awkward proximity. Such displacement required that a different set of qualities be identified as the most significant for purposes of comparison. Most frequently, evidence from the behavioral or moral sphere replaced the merely anatomical. Thus, throughout the nineteenth century naturalists debated the rival claims of dogs and apes to be top animal, and therefore closest to humankind. In 1881, for example, George J. Romanes, a close friend and colleague of Darwin's with a special interest in animal behavior, celebrated the "high intelligence" and "gregarious instincts" of the dog, which, he claimed, gave it a more "massive as well as more complex" psychology than any member of the monkey family possessed.¹⁰ And since the competing closeness so constructed was clearly figurative, the whole animal creation was thereby implicitly removed to a more comfortable distance.

From this perspective, the creatures who approached human beings most closely were likely to be the domesticated pets and livestock who shared their lives, the wild creatures whom they routinely hunted, or, in a more attenuated relationship, the animals, wild or tame, who served as traditional metaphors for human attributes. And although, as a rule, these alternative arrangements were not embodied in abstract schema, they did reflect persistent conventions in the graphic portrayal of people and animals. The notion that similar appearance indicated similar character had a long history. As an English translation of the seventeenth-century French artist Charles LeBrun, who himself consolidated and elaborated a much older tradition of linking human attributes with those of beasts, noted, "the Physiognomists say that if a Man happens to have any part of his Body resembling that of a Brute, we may . . . draw Conjectures of his Inclinations."¹¹ LeBrun's mechanistic deconstruction of personality traits doubled as a recipe for depicting them. He proposed, for example, that the point of intersection between a line drawn from the corner of the line across the upper eyelid and a line drawn from the nose indicated mental power: "when these two lines meet in the Forehead, it is a sign of Sagacity, as . . . in the Elephants, Camels, and Apes: But if the

Angle meets upon the Nose, it shews Stupidity and Weakness, as in Asses and Sheep."[12] Humans could follow the pattern set by any of these creatures.

And although the authority of LeBrun's analyses waned along with the assumption that the essential nature of each individual and each species was so obvious and so invariant, English editions of his work continued to appear well into the nineteenth century. Images evoking the reciprocal resemblance between humans and members of a range of other species constituted part of the stock visual repertory of both humorous and sentimental social commentary, figuring in genres from ephemeral satire to academic painting. Thus a mid-Victorian lithograph displayed a crammed panorama of natural history buffs, whose physical appearance unflatteringly reflected the adjacent objects of their fascination, while *Punch* presented a gentleman with a pendulous belly and a beaky nose gazing with embarrassed recognition at a penguin. So conventional was this mirroring that, as the many anthropomorphic canvases of Edwin Landseer illustrated, its evocation required the presence of only a single member of the pair (usually the nonhuman one).

These juxtapositions matched people with a dizzying variety of zoological doubles, implicitly constructing polymorphous, eccentric, and inconsistent patterns of relationship. And artists could establish unconventional countertaxonomies upon even the most profoundly scientific basis. When the late-eighteenth-century painter Sawrey Gilpin claimed that "The lines, w[hich] form y. countenance of y. lion approach nearer to those of y. human countenance, than y. lines of any animal with w[hich] we are acquainted," he may have been influenced as much by the animal's traditional role as the "King of y. beasts" and by analogy in the style of LeBrun, as by any technical analysis.[13] Only a few years later, however, the romantic artist Benjamin Haydon based his assertion of the same special relationship on anatomical research: "while dissecting a lion, . . . I was amazingly impressed with its similarity as well as its difference in muscular and bone construction to the human figure. It was evident that the lion was but a modification of the human being."[14]

The final anatomical work undertaken by George Stubbs was entitled *A Comparative Anatomical Exposition of the Human Body with that of a Tiger and a Common Fowl*. In it Stubbs too expressed an alternative vision of the human context in conventional anatomical terms. The mere juxtaposition of these three creatures suggested relationships that were usually unnoticed or nonexistent, and Stubbs's aggressively realistic (or hyperrealistic) depictions nevertheless exaggerated the unusual resemblances evoked by his selection. The meticulously detailed plates showed primate, carnivore, and gallinaceous

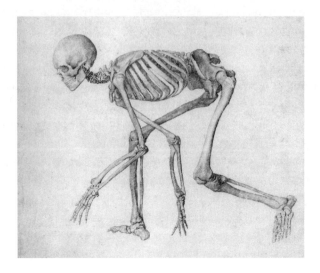

Human Skeleton, Lateral View, in Crouching Posture, by George Stubbs.

bird in more or less identical positions and in identical stages of dissection, from the intact body, through the successive removal of skin, muscles, and internal organs, to the naked bones. Not only did the human model assume quadrupedal postures, but he was portrayed with a skeletal configuration—relatively narrow shoulders and wide hips—that made him resemble the chicken more closely than might have been expected.[15]

In a similar vein the author of a Victorian anatomy book designed for artists, in which creatures were shown first with their skin and flesh, and then without them, noted that the juxtaposed skeletons of a dancing bear and a human bear warden illustrated that "the bears (genus Ursidae) have a claim superior to that of apes and monkeys for the nearest proximity to human beings, on account of their plantigrade feet and . . . erect attitude."[16] And in 1891 the director of the British Museum (Natural History) used a drawing of a human skeleton standing next to an equine one, with their corresponding joints anthropomorphically and vernacularly labeled as "wrist," "knee," "ankle," and so forth, as the frontispiece for a monograph entitled *The Horse: A Study in Natural History.*[17] The exhibit on which it was based, actual skeletons standing side by side, was the first thing encountered by visitors to the museum's large collection of skeletal equids, which was otherwise classified according to contemporary zoological consensus.[18] The juxtaposition implicitly presented the social connection between the highbred horse and his rider or trainer as at least equivalent to the anatomical or phylogenetic connection between the horse and the zebras and asses that constituted the rest of the exhibit.

Such taxonomic and countertaxonomic relationships both represent and determine our notions about our zoological place. The category "beasts" has never been either homogeneous or stable. Rejecting one kind of animal connection has often opened the way for another. Even a brief survey of Victorian ideas suggests the range of possible relationships to other creatures. And a deep acknowledgment of similarity remains as firmly embedded in contemporary culture as does the scientific or theological assertion of difference.

Notes

1. Tyson, *Orang-outang*, preface, n.p.
2. *Historical Miscellany*, 3:288–89.
3. Linnaeus, *Systema Naturae*, 20–24.
4. Lydekker, *Hand-book to the Carnivora*, 45–48; poster in the Cambridgeshire Collection, Cambridge Central Library.
5. Even Darwin accepted the reports of these hybrids, albeit hesitantly, and was relieved when they were discredited. Darwin, *Variation of Animals and Plants under Domestication*, 2:447.
6. William Lawrence, *Lectures on Comparative Anatomy*, 127, 131.
7. See, for example, Richard Owen, "On the Anthropoid Apes—and their relations to Man," *Proceedings of the Royal Institution of Great Britain* 2 (1855): 41.
8. [Adam Sedgwick], "Objections to Mr. Darwins Theory of the Origin of Species," *Spectator*, 14 March and 7 April 1860, reprinted in Hull, *Darwin and His Critics*, 164–65.
9. Darwin, *Descent of Man*, 919.
10. Romanes, *Animal Intelligence*, 439.
11. Le Brun, *Conference*, 40; Montagu, *Expression of the Passions*, 20–24.
12. Le Brun, *Conference*, 46.
13. Gilpin, "On the character and expression of Animals," 5, 13.
14. Haydon, *Lectures on Painting and Design*, 1:13.
15. Stubbs, *Anatomical Works*, 111–279
16. Hawkins, *Comparative Anatomy*, 47.
17. Flower, *Horse*, frontispiece.
18. Lydekker, *Guide to the Specimens of the Horse Family*, 4.

— 12 —

Counting Sheep in the English Lake District

Rare Breeds, Local Knowledge, and Environmental History

In 2001, British television viewers were horrified to witness an apparent military assault on the nation's ovine population. Civilian resources had proved inadequate to contain an outbreak of foot-and-mouth disease, and so the army was called in to expedite the destruction and disposal of sick animals as well as those considered at risk, which included apparently healthy herds and flocks living within a mile or so of any actual infection.[1] This dramatic episode had serious political and economic implications. The British livestock industry was just beginning to recover from the protracted crisis caused by BSE (bovine spongiform encephalopathy, or mad cow disease). That earlier crisis had been significantly intensified by misunderstanding and mismanagement on the part of the conservative governments in power during the 1980s and 1990s, and so there was a great deal of curiosity and apprehension about the way Tony Blair and New Labor would deal with this challenge. From a strictly medical or veterinary perspective foot-and-mouth disease is much less serious than BSE. It is an extremely contagious viral infection, affecting a range of hoofed animals, and a few unhoofed ones such as elephants. The major symptoms are blisters and sores around the mouth and feet, and in some cases elsewhere on the body. Afflicted animals are usually

---

"Counting Sheep in the English Lake District: Rare Breeds, Local Knowledge, and Environmental History" was presented at "Animals in History: Studying the Not So Human Past," a conference held at the Literaturhaus in Cologne, Germany, cosponsored by the Angloamerican Institute of the University of Cologne and the German Historical Institute, 18–21 May 2005.

sick for several weeks, during which time they are reluctant to eat, and recovery can be slow. It is not usually fatal, except to young animals, but the disease can cause a serious reduction in meat and milk production.[2] Even worse, from a financial perspective, is the quarantine that threatens the animal products of any country normally free of foot-and-mouth disease that harbors an outbreak. To deal with this second catastrophe looming over the British livestock industry, the government decided, figuratively speaking, to launch the nuclear option.

It is clear that the drastic eradication policy was based on economic rather than medical concerns. Foot-and-mouth disease did not pose a serious threat to human health—that is, people who eat meat or milk from infected animals do not become sick, although occasionally people who work closely with infected animals contract a mild "flu-like" illness. The sheep, cattle, and pigs, who, sick or not, were the most numerous victims of the outbreak, died at the hands of the British government, not from the foot-and-mouth virus. (From their point of view, it is hard to know whether being shot in their home fields was a better or worse experience than the one that otherwise awaited many of them in a slaughterhouse.) But the farmers who lost their livestock nevertheless had many human companions in suffering. Although outbreaks were scattered over much of the United Kingdom, they were concentrated in a few regions. Of the total cases 45 percent occurred in Cumbria, which occupies the northwesternmost corner of England; in that county alone, more than 1 million sheep were slaughtered, along with more than 250,000 cattle, pigs, goats, and deer.[3] The livestock population of Cumbria thus dwarfs its human population of only half a million, less than 1 percent of the total population of the United Kingdom. Farmers were not the only local residents to mourn the mass slaughter of their animals. Of the fifty-one contributors to a collection of reminiscences of the outbreak entitled *Foot and Mouth—Heart and Soul,* only six identified themselves as farmers. The rest had occupations such as veterinarian, slaughterman, hotelier, mountaineer, broadcaster, cartoonist, and redundant sheepdog.[4] And local residents were not the only people with a proprietary fondness for the Cumbrian hills and valleys. The videos that showed soldiers in battle fatigues beside flaming pyres of carcasses also featured some of Britain's most celebrated country landscapes.

Columns of black smoke are inconsistent with conventional notions of rural beauty, and so this kind of free publicity had a predictably dampening effect on Lake District tourism. Perhaps more important, quarantine regulations imposed to prevent the spread of foot-and-mouth disease via shoes and clothing meant that walkers were largely confined to paved public roads, and prohibited from tramping across fields and fells. Although advertisements announced that the Lake District was open, this claim turned out to refer primarily to tea shops and pubs. Even the gardens and parks of many large country houses were off limits; for the duration of the emergency admission fees only bought access to the facades and the furniture. (That these measures were likely to produce inconvenience rather than any real effect was obvious even after the most casual reflection. The pools of disinfectant through which travelers were required to walk and drive were shallow and infrequently refilled; the signs barring access to the fells would not discourage determined walkers; and the uplands were the home of many animals besides livestock who were equally capable of spreading infection—especially if they snacked on the piles of dead sheep that were sometimes left to fester after the soldiers had done their work.)

If potential tourists could express their unhappiness with this state of affairs by staying away, local residents had to endure the combined calamities on the spot, sometimes even confined to their farms along with their doomed animals. A study conducted at the Institute for Health Research of nearby Lancaster University concluded that many livestock owners suffered emotional traumas that significantly transcended their material losses.[5] (Of course these traumas were not limited to farmers in Cumbria. Those in other affected areas experienced similarly intense suffering, as is shown by the heightened emotionalism of the headlines in a special tabloid supplement documenting the foot-and-mouth outbreak in the rural northeast of England: "The Pain and Fear of Foot and Mouth," "Apocalypse Enfolds Traumatised Shires," "Grim Summer Punctuated by Hotspots of Despair," and "Lambs to the Slaughter as Chaos Grips the Industry."[6])

Dramatic though they were, however, most of the effects of the foot-and-mouth crisis proved to be temporary. The outbreak seemed to be under control by the autumn of 2001, despite widespread dissatisfaction (to put it very mildly) with the government response.[7] By the next tourist season both sheep and walkers had returned to the Cumbrian fells. As had been the case with the earlier BSE crisis, public hysteria and public memory gradually receded, along with the frequency of alarming newspaper headlines and television reports. But at least one consequence of the draconian (if belated) response to the outbreak threatened to be permanent.

Although most livestock animals in Great Britain, as elsewhere, belong to a few favored modern breeds, the island also hosts a selection of minority breeds, living reminders of the rich British traditions of animal husbandry and agricultural improvement. According to the Rare Breeds Survival Trust, the organization that catalogues and monitors such breeds, those most at risk of disappearing—for example, Irish moiled cattle or Cleveland bay horses—may number only a few hundred individuals. They are likely to be concentrated in a few localities, or even on a few farms, which means that either a fast-moving epizootic or a vigorous prophylactic cull could decimate a breed, or wipe it out completely.[8] One of the strategies for ensuring the continuation of breeds designated as endangered, therefore, is to avoid geographical concentration by establishing herds or flocks at widely separated locations. (Given the relatively small size of Great Britain and the encompassing nature of modern webs of animal transportation, this strategy may merely reflect wishful thinking; in 2001, for example, foot-and-mouth disease occurred almost everywhere that there were large livestock populations.)

The dominant sheep breed in the Lake District never appeared on the watchlist of the Rare Breeds Survival Trust, since, neither at the beginning of the foot-and-mouth outbreak, nor, as it turned out, at the end, did its numbers approach the 3,000–individual maximum for the least threatened category. But the 100,000–strong Herdwick flock was nevertheless at similar risk from the policy of quarantine and, especially, aggressive culling, since the animals were concentrated in a relatively small region (approximately thirty miles square) where foot-and-mouth disease was especially prevalent, and where, in consequence, military maneuvers were particularly intense. The beleaguered Herdwicks therefore featured prominently in the press coverage of the foot-and-mouth outbreak; they were even singled out for special concern in a report that focused on the impact of the disease in the southwest of England, hundreds of miles distant.[9] To some extent this simply reflected the enormous magnitude of the cull. But quantification, whether in terms of animal corpses or financial costs, could not adequately express either the extent or the nature of the feared loss. The Herdwicks were valued for a set of intrinsic qualities that distinguished (or seemed to distinguish) them from other breeds. Further, it was widely predicted that their disappearance would irremediably disrupt the Cumbrian landscape and economy, producing a gap that no fellow sheep could ever fill.

Modern Herdwick aficionados emphasize the hardiness of their favorite

"Herdwick Rams," from Frank Wilson, *Westmorland Agriculture*, 1912.

sheep in the face of harsh climatic conditions and their ability to sustain themselves by foraging on sparse upland vegetation. The National Trust, which owns about a quarter of the Lake District and therefore also a great many Herdwicks, praises the flavor of their meat, urging visitors to its tea shops not to "miss the opportunity to taste Herdwick meat—a speciality of the Lake District!" and offering all-Herdwick heritage meat hampers for £75 and £90. The Trust also provides a market for Herdwick wool, which is too coarse for clothing, using it to produce a line of heritage carpets.[10] (An artificial boost to the market was needed because the price of Herdwick fleeces had dropped so low that it had become unprofitable to farm them—a different anthropogenic threat to the breed's survival.) In these characteristics, as well as in the breed standards specified by the Herdwick Sheep Breeders' Association, the animals do not seem to have altered much over the past several centuries.[11] In 1805, for example, an agricultural survey of the north of England described the Herdwicks as "lively little animals, well adapted to seek their food amongst these rocky mountains"; in 1837 the veterinarian William Youatt declared that "the principal value of this breed is its hardiness."[12] One mid-nineteenth-century agricultural pundit characterized them as "small, well made, active, and polled [that is, hornless—current breed standards require that rams be horned], the faces and legs being more or less mottled with black [color is a particularly volatile characteristic—current breed standards identify such mottling as an 'objection']."[13]

Despite this long history of separate recognition, however, in their appearance, their virtues, and their habits, Herdwicks so strongly resemble the sheep breeds that emerged or were developed in other, topographically similar regions of Britain, that they are often treated as varieties of a single type. For example, when the Ministry of Agriculture and Fisheries compiled a catalogue of British livestock breeds in 1927, it included the Herdwick (along with the Scotch Blackface, the Lonk, the Derbyshire Gritstone, the Rough Fell, the Swaledale, the Limestone and Penistone, the Cheviot, the Welsh Mountain, the Radnor, the Exmoor Horn, and the Dartmoor—most of which bear the names of their rugged homelands) under the rubric "mountain sheep . . . distinguished by their hardiness and ability to thrive on poor food distributed over very large areas, their activity, small size, and extremely high quality of mutton."[14] A more recent survey groups the same breeds as "sheep of hill and mountain."[15]

It is possible to understand the impulse to distinguish the Herdwick on the basis of not much difference as an expression of more general tendencies in eighteenth- and nineteenth-century livestock breeding. This was the period in which modern breeds of cattle, pigs, and sheep—that is, breeds based on descent rather than on function or geography—were established, and then improved (or reified) through the institutions of societies, shows, and books of pedigrees.[16] The Herdwicks basked in most of these forms of attention. Many of the early descriptions of the breed allude to a shadowy attempt (whose time, place, and agents were never specified) to corner the reproductive market, according to the pattern established by Robert Bakewell and the Dishley Society with regard to New Leicester sheep (the major livestock-marketing success story of the late eighteenth century). Youatt, for example, described a nameless association "one of the regulations of which was, that they never should sell a ram, and not more than five ewe-lambs in one season." He was relieved to report that unspecified means had been found "to elude this illiberal and shameful monopoly."[17] Much better documented were the prizes for Herdwicks offered at local shows, beginning with that of the Penrith Agricultural Society in 1833. The first national prizes were offered at the 1855 meeting of the Royal Agricultural Society, which was held in nearby Carlisle. Although no Herdwick flock book was compiled until well into the twentieth century, the West Cumberland Fell Dales Sheep Association was founded in 1844 in order to improve and publicize the breed, which it accomplished mainly by holding annual shows. By 1910, the association was offering prizes for Herdwicks in twenty-one separate classes (segregated mostly by age and sex, but sometimes by such characteristics as color of fleece and face).[18] The strong sense of breed produced by all of these efforts increased the visibility and prestige of the Herdwick designation (branding, in modern terms), and therefore the cash value of the individual animals within it.

The farmers and fanciers who deployed such means to enhance breed recognition and prestige normally did not present their agenda in these crudely commercial or quantitative terms. Instead, they characterized themselves as protectors of the breed or advocates of its interests. They frequently claimed that they were working to defend their chosen breed's purity, which implied the existence of deep historical roots and long reproductive isolation (claims and implications that were at odds with the demonstrably recent consolidation of most breeds). For example, an article in the preeminent Victorian agricultural journal asserted that the Herdwicks possessed "more of the characters of an original race than any other in the county" and that they showed "no marks of kindred with any other race."[19] This assertion was buttressed by the breed's strong association with the hills of Cumberland and Westmorland—as well as by the folk etymology enshrined in the *Oxford English Dictionary*, which defines Herdwicks as "a hardy breed of mountain sheep in Cumberland and Westmorland. Supposed to have originated on the herdwicks of the Abbey of Furness"—as a common noun, "herdwick" is an obsolete term for the tract of land supervised by an individual shepherd; Furness Abbey lies on the southern coast of Cumbria.

Such claims of originality or indigenousness were, however, simultaneously undermined by numerous Herdwick origin stories that agreed on nothing but the fact that the sheep had arrived in Cumbria by ship. According to Youatt, they were descended from a group of Scottish sheep traveling aboard a ship that stranded on the Cumberland coast early in the eighteenth century. In this version, the ship's captain sold the sheep to local farmers, who set them free to roam around on the hills.[20] Alternative accounts were less matter-of-fact and still less verifiable, deriving the breed from the wreck of the Spanish Armada or of some more ancient and nebulous Scandinavian voyagers.[21] Modern historians of livestock tend to discount all these romantic stories, instead stressing the telltale physical similarities that link Herdwicks to the other hardy mountain sheep of Wales and the west of England.[22] Nevertheless, the claims of exotic origin resurfaced persistently in press accounts of foot-and-mouth disease, along with (somewhat inconsistent) assertions that the breed was native to the Lake District. Thus one credulous journalist confidently credited the Vikings with their introduction, while also citing a local farmer as authority for their indigenous status.[23]

Strictly speaking there are no indigenous British sheep, since domesticated European sheep all descend from the mouflon of the eastern Mediterranean. A few British breeds, such as the Soay (listed by the Rare Breeds Survival Trust as "vulnerable," meaning that there are fewer than nine hundred individuals) and the Shetland, resemble the animals that were originally introduced to Britain between two thousand and four thousand years ago, but the Herdwick is not among them.[24] Their indigenousness nevertheless remains a highly valued attribute of the Herdwicks, albeit a somewhat elusive and unstable one. And if the breed could not claim a truly chthonic relationship to the Cumbrian hills, it did have several powerful connections to the landscape it inhabited. Not only were Herdwicks strikingly well adapted, through a combination of natural and artificial selection, to the rigorous living conditions on the northern fells, but they were widely believed to possess unique local knowledge. As one agricultural encyclopedia put it, "when wandering uncontrolled over the mountains, they display remarkable instinctive sagacity . . . no sheep could . . . be better fitted for the locality."[25]

Traditional descriptions of Herdwicks often emphasized the special affinity of each flock for its particular territory, asserting that information about its nooks and crannies—where best to feed and shelter, where the footing was dangerous—was passed down through the ovine generations, without human assistance or interference. In this case, familiarity appeared to breed attachment rather than contempt. The fondness of Herdwick sheep for their native "heaf" (a northern dialect term for their accustomed pastures) was so widely recognized, that special legal clauses taking account of their tendency to return to what they considered their home were often inserted into contracts for their purchase or letting; when a tenant took over a new farm, the resident flock often constituted part of the bargain.[26] Thus, when the entire breed was threatened by foot-and-mouth disease, a cultural as well as a genetic heritage seemed on the verge of disappearing. An article in the *Independent*, which characterized the sheep themselves as one of the most "innate" attractions of the Lake District, explained that the "'hefting' instincts" that

A Lake District scene: *Grange in Borrowdale: Early Morning*, from W. T. Palmer, *The English Lakes*, 1908.

were "passed from ewe to lamb" allowed the sheep to roam up to forty miles without getting lost. Ironically, as it happened, the same tendency put them into contact with the largest possible number of potential carriers of foot-and-mouth disease and made them especially likely to transmit the infection onward.[27]

The reporter went on to speculate that the sheep were not the only feature of the Lake District that might disappear as a result of the outbreak. He further feared that the whole ecology of the region might be changed "beyond recognition."[28] Although this concern sounds exaggerated, it was well founded (or would be well founded if the Herdwicks turned out to be truly irreplaceable—that is, if no introduced sheep breed could adapt with equal effectiveness to this challenging terrain). Ever since the Lake District emerged as an iconic English setting in the late eighteenth century, it has been prized as a profoundly natural landscape—if not unpeopled, then essentially unaltered by human activity. The attractiveness and durability of this characterization have proved impressively invulnerable to modification by counterevidence, no matter how robust. The poet William Wordsworth, who wrote a *Guide to the Lakes* for visitors (first published in 1810 and frequently reissued), surveyed the human impact on the Lake District beginning with the ancient Britons. But at the same time that he described repeated anthropogenic alterations of the landscape, he implicitly erased them. Especially as his gaze descended the social ladder, he conflated human and natural processes, so that the cottages of the humbler inhabitants reminded him "of a production of nature, and may . . . rather be said to have grown than to have been erected; . . . so little is there in them of formality, such is their wildness and beauty."[29] Following the path blazed by Wordsworth, the famous opium eater Thomas DeQuincey called the Lake District "one paradise of virgin beauty."[30]

In the course of the nineteenth century, human impact on the Lake District became increasingly obvious, largely as a result of the synergistic impact of tourism and railway construction. None of this diminished the appeal of Wordsworth's perspective to subsequent admirers of what they perceived as the unspoiled landscape. On the contrary, the allegedly virgin or natural condition of the Lake District seemed to become more appealing as the evidence supporting it diminished. This is not to say that such evidence had ever been very strong; modern transportation technology was far from uniquely transformative. The region had been shaped for more than five millennia by human farmers, miners, and pastoralists.[31] But the distinctive bare, stark upland scenery that was the focus of the most intense admiration was only indirectly a human creation. Norse invaders settled in Cumbria during the tenth century, clearing woods to make farmsteads, and setting their cattle and sheep loose to roam the hills. Over time, the more efficient and less choosy sheep became the dominant ungulate presence, steadily nibbling away at struggling tree saplings and so preventing reforestation.[32] Whether these were the storied ancestral Herdwicks or whether the Herdwicks subsequently replaced them, the Viking chattels and their successors were indeed responsible for producing and maintaining the ecology that continues to characterize the higher altitudes of the Lake District, as well as its distinctive appearance. The floral and faunal assemblages that coexisted with them on the denuded fells were very different from those that had flourished in the preexisting forests.[33]

With regard to mammals, at the end of the nineteenth century this fauna included foxes, badgers, otters, pine martens, stoats, polecats, several species of deer, and a varied mix of rodents, insectivores, and bats. Because of the region's remoteness from centers of human population and its unsuitability for intensive development, remnant populations of some species lingered there after they had become extinct in most other parts of Britain. Isolation was no guarantee of survival, however, especially since local human residents, though not numerous, were themselves enthusiastic predators. Cumbrian wolves, bears, wild boars, and beavers had long since disappeared, along with their lowland conspecifics; and in 1892 the naturalist H. A. Macpherson lamented that even the memory of the wild cat (which resembles a particularly formidable tabby) "no longer survives among the venerable dalesmen whose grandfathers were the chief instruments of its extinction."[34] He further noted that badgers, which had been plentiful throughout most of the eighteenth century, had vanished by 1875; that the numbers of pine martens "have of late years greatly decreased"; that "within the last thirty years the Polecat has become very scarce in Lakeland"; and that otters were threatened "by those who surreptitiously trap these fine animals."[35]

Although Macpherson offered a single explanation (that is, human hunting) for all these declines, a century later the opposed pressures of environmental change and habitat preservation had produced a range of results. Badger populations had rebounded, while the otter and the pine marten tottered on the verge of local extinction (both subsequently saved by recolonization from outside), and the polecat had toppled over.[36] Nor were people, whether in the guise of *Homo venaticus* or of *Homo economus,* the only source of challenge to Cumbrian wildlife. Macpherson's catalogue included only one kind of squirrel, the indigenous red *Sciurus vulgaris,* but a 1970 survey noted of *Sciurus carolinensis,* the invading grey squirrel of North America, "Stragglers reported since World War II . . . Profoundly hoped not yet breeding." These hopes were doomed to disappointment, since gray squirrels have established themselves in several parts of the Lake District, and seem likely to outcompete the red squirrels there, as they have done elsewhere in Britain, despite the vigorous legal protection of the native squirrel.[37]

In its doomed struggle to resist the transatlantic onslaught the red squirrel has sometimes served as a national symbol of embattled purity and isolation. For example, when the Heritage Lottery Fund awarded £626,000 to protect what supporters affectionately termed "the real Squirrel Nutkin," the grant drew praise from the entire political spectrum, not just the mainstream parliamentary parties, but also, for example, the (self-described) "patriotic nationalist" British National Party.[38] This generalized iconic status has not translated to the Lake District, however, even though the area remains one of its few remaining refuges, and even though the red squirrel's prominent place in British affections owes much to the work of Beatrix Potter, herself a celebrated Lake District resident and militant landscape preservationist. Like the other wild animals that inhabit the hills and valleys of the Lake District, it is not specifically Cumbrian, but rather part of a fauna that was once widely distributed in Britain and, indeed, throughout northern Europe (which is why dwindling otter and pine marten populations could be successfully reinforced by imported animals). The most distinctive feature of Lake District natural history is the alpine floral assemblage of the high fells, the residuum of the vegetation that colonized the entire area after the retreat of the last Pleistocene glaciation; it is unique in England, although similar to that of the high altitudes of Scotland and Wales (as well as to that of tundra areas much farther north).[39]

Large ubiquitous animals tend to have more charisma than small and relatively inaccessible plants, however attractive and indigenous. And so, for that reason alone, it is not surprising that the Herdwicks have emerged as popular symbols of the Lake District, rather than the alpine flowers, despite their recent accession to indigenous status and their incontestable status as domesticates. For those who, like Thomas DeQuincey, romantically view the Lake District as a preserve of unspoiled nature, these characteristics may emphasize the unsuitability of the Herdwicks for their representative role. For those with a more historically nuanced understanding of the Cumbrian landscape, the recent advent of the Herdwicks and their liminal lifestyle may actually enhance their iconic qualifications. Whether technically indigenous or not, the Herdwicks have strong physical and economic ties to their place; they live there and (almost) nowhere else; they have helped to produce its modern shape. Further, like the landscape itself, they seem wilder than they are; that is, they appear to be independent and free ranging, but their lives (and, indeed, their very existence) are ultimately determined by human economic exigencies. Their liberty is intermittently interrupted so that their owners can shear, cull, or otherwise manage them. They are both accessible (that is, there are a lot of them and they are everywhere, not only in the fields, but grazing and napping beside the roads and even on top of them) and also inaccessible (that is, they are skittish, and tend to retreat when approached).

One other domesticated animal—the border collie—shares many of the characteristics that make the Herdwicks iconic. They roam equally widely over the hills, and often, in a sense, more wildly. Unconstrained sheep behave very similarly to constrained sheep, confining most of their attention to what-

"The English Sheep-Dog," from John Walsh, *The Dog in Health and Disease,* 1879.

ever they are trying to eat, while unconstrained off-duty collies may undergo what appears to be a Jekyll-and-Hyde transformation, persecuting members of other flocks when they are not responsible for herding their own. Paradoxically, however, this apparently uncharacteristic behavior underlines the extent to which the dogs belong within the human orbit. Sheep worrying is an extreme and (from the human perspective) unfortunate expression of the same intelligence and aggression that make collies such effective shepherds. If not firmly controlled by their masters, they would play a very different role in the web of Cumbrian life (or perhaps no role at all, since dogs caught in the act of harassing sheep are apt to be shot).

The foot-and-mouth epizootic of 2001 (or the official response to the outbreak and the subsequent journalistic accounts) cast a lurid and flickering light on the relationship of Herdwick sheep to their natal hills—a relationship that, if not primordial, has deep historical roots and powerful cultural resonance. The episode is of obvious interest to historians of agriculture and of public health, whether of humans or of other species. It is also of interest to environmental historians. In part this is because any story that unfolds in a rural landscape, especially one that has experienced or is vulnerable to significant change, engages concerns at the core of environmental history. But it is not only the transformation of the Lake District landscape—either temporarily, by its conversion into a site of guerrilla warfare, or, potentially, permanently, by the removal or replacement of a large component of its mammalian population—that constitutes the importance of the foot-and-mouth-disease epidemic of 2001, along with its antecedents and contexts, for environmental historians. Every component of this story engages the ambiguous boundary (or nonboundary) between the wild and the domesticated, the natural and the human: not only the reengineered landscape and the untrammeled sheep, but the responses of breeders, farmers, government officials, journalists, and members of the public, both over time and to the immediate crisis. As one thoughtful commentator put it at the time, "foot-and-mouth has shown . . . that sheep . . . provide a vital connection linking people, animals, the land and capital."[40]

Environmental history is relatively new as historical subdisciplines go, and in North America it is rooted in the history of the frontier and of European encounters with what has conventionally, although far from unproblematically, been perceived as wilderness.[41] As a result, many researchers have tended to emphasize relatively (or at least apparently) static features of the landscape of western North America, such as mountains, rivers, and forests. If animals have figured in such accounts it has most often been as the targets or rivals of hunters, or, in a more literary register, as the symbolic representatives of nature.[42] In Europe, environmental history is even newer (that is, the label is even newer), but its roots are more complex and at least equally deep. They encompass not only the study of places perceived as wild, whether on the imperial frontiers or on the undeveloped peripheries of Europe, but also the study of the long tamed and deeply familiar countryside. Increasingly, however, even American environmental historians have turned their attention to the landscapes of domestication that replaced many former frontiers. Inevitably, animals have played a more prominent role in the complementary histories of environments transformed by European agricultural practices, from the subsistence farms of eastern New England to the ranches of central Mexico.[43]

If the agricultural landscape, especially the newly agricultural landscape, can now be understood as a kind of border territory, a contact zone where the natural and the artificial or technological intermingle, domesticated animals have long occupied a similarly liminal position. Their simple presence inevitably brings their absent wild relatives to mind. One reason that few ordinary people kept pets for modern reasons—that is, to provide pleasure rather than labor—until late in the eighteenth century, was that even the smallest and least ferocious dog or cat seemed to represent the intrusion of wild and threatening nature into the family circle; as technology made nature seem less terrifying, animals could become explicit friends rather than ostensible servants.[44] Charles Darwin played on the analogy between domesticated animals and wild animals at the beginning of *On the Origin of Species*, using artificial selection, the technique by means of which breeders tried to determine the characteristics of pedigreed livestock and pets, to introduce the much more radical and profound notion of natural selection, the process by means of which species evolve in nature.[45] The analogy has been continually strengthened by the practices of both scientific taxonomists primarily concerned with wild animals and breeders (and others) primarily concerned with domesticated animals. Thus domesticates are routinely assigned to a different species than that occupied by their wild progenitors, even if, as is the case with the dog, those relatives still exist and interbreed freely given the opportunity.[46] Conversely, the Rare Breeds Survival Trust has borrowed the categories developed for vanishing wild species, even though much less is at stake in terms of biodiversity and ecology if a single strain of cattle or sheep disappears. Perhaps as a tacit acknowledgment of this difference of degree, the RBST has added "traditional," its least threatened category, to "critical," "endangered," "vulnerable," and "at risk."[47]

The understandings produced by these layered connections are often submerged, but at the time of the foot-and-mouth outbreak they surfaced dramatically. Or, to put it another way, figurative relationships suddenly seemed very concrete. The armed assault on the Herdwick sheep represented an attack on both the domesticated countryside and the unspoiled natural landscape. It connected the suffering of individual animals with the suffering of their human proprietors and with the less acute deprivation of the larger public for whom the Cumbrian hill country was both a popular recreational destination and a sacred national space. The Herdwicks were demonstrably neither native nor wild, both qualities that figured prominently in the Lake District mythos. Nevertheless, they still seemed, as they had two centuries earlier, "peculiar to that high, exposed, rocky, mountainous district."[48] And despite their murky origins and their intermediate position between human husbandry and the harsh life of the fells—or perhaps because of them—this hardy and home-loving breed of sheep had become the vehicle through which deep understandings of their environment were routinely expressed.

Notes

1. An elaborate verbal and visual record of the outbreak is available on the BBC website, http://news.bbc.co.uk/1/hi/in_depth/uk/2001/foot_and_mouth/default.stm. For a historical overview, see Woods, *Manufactured Plague*.
2. See the World Organisation for Animal Health website, http://www.oie.int/eng/maladies/fiches/A_A010.htm.
3. Convery et al., "Death in the Wrong Place?" 103–4.
4. Caz Graham, *Foot and Mouth—Heart and Soul*.
5. Convery et al., "Death in the Wrong Place?" 107. See also Franklin, "Sheep-watching."
6. "Foot and Mouth: How the North Lived through a Nightmare," supplement to *Journal* (Newcastle), 21 January 2002, 3, 7, 16, 18.
7. See, for example, *Inquiry Report*, esp. 51–56.
8. Lutwyche et al., "Special"; Rare Breeds Survival Trust website, http://www.rbst.org.uk/watch-list/main.php.
9. Gripaios et al., "Economic Impact of Foot and Mouth Disease"; Lawrence Alderson, "Foot-and-Mouth Disease."
10. See the National Trust website, http://www.nationaltrust.org.uk/main/.
11. See the Herdwick Sheep Breeders' Association website, http://www.herdwick-sheep.com/herdwick_standards/index.htm.
12. Bailey and Culley, *General View*, 246; Youatt, *Sheep*, 279.
13. Martin, "Sheep," 109.
14. United Kingdom, Ministry of Agriculture and Fisheries, *British Breeds of Livestock*, 93.
15. Hall and Clutton-Brock, *Two Hundred Years of British Farm Livestock*, 111.
16. For extended discussions of these developments, see Ritvo, *Animal Estate*, 45–141; "Possessing Mother Nature: Genetic Capital in Eighteenth-Century Britain," chapter 10 in this volume; and Ritvo, *Platypus and the Mermaid*, 104–20.
17. Youatt, *Sheep*, 279.
18. Frank Wilson, *Westmorland Agriculture*, 161–63.
19. Dickinson, "On the Farming of Cumberland," 264.
20. Youatt, *Sheep*, 278.
21. Frank Wilson, *Westmorland Agriculture*, 151.
22. Hall and Clutton-Brock, *Two Hundred Years of British Farm Livestock*, 116; Trow-Smith, *History of British Livestock Husbandry*, 134–35.
23. *Express*, 27 March 2001, 8.
24. Clutton-Brock, *Natural History of Domesticated Animals*, 53; Hall and Clutton-Brock, *Two Hundred Years of British Farm Livestock*, 99.
25. Martin, "Sheep," 109.
26. Frank Wilson, *Westmorland Agriculture*, 159; Bruce Thompson, *Lake District*, 177.

27. Ian Herbert, "Foot and Mouth Crisis: Cumbria," *Independent,* 27 March 2001, 5.
28. Ibid.
29. Wordsworth, *Illustrated Wordsworth's Guide to the Lakes,* 105.
30. DeQuincey, *Literary Reminiscences,* 311.
31. For an overview of the development of the Cumbrian landscape, see Pearsall and Pennington, *Lake District.*
32. Fell, *Early Settlement in the Lake Counties,* 84; Winchester, *Harvest of the Hills,* 103–4.
33. Simmons, *Moorlands of England and Wales,* 24.
34. Macpherson, *Vertebrate Fauna,* 9.
35. Ibid., 44, 25, 27, 35.
36. Delap, "Mammals"; Webster, "Mammals in Cumbria."
37. Macpherson, *Vertebrate Fauna,* 76; Delap, "Mammals," 193; Webster, "Mammals in Cumbria," 84.
38. Brian Dooks, "Lottery cash aids efforts to save the red squirrel," *Yorkshire Post,* 11 April 2006; British National Party website, http://www.bnp.org.uk/index.php.
39. Hervey, "Flowering Plants, Ferns and Mosses."
40. Franklin, "Sheepwatching," 4.
41. For an influential explication of these problems, see Cronon, "Trouble with Wilderness."
42. Karl Jacoby has examined conflicting understandings of the role of wild animals in nineteenth-century American frontiers in *Crimes against Nature.*
43. Among many sources, see Crosby, *Columbian Exchange;* idem, *Ecological Imperialism;* Cronon, *Changes in the Land;* Melville, *Plague of Sheep;* and Virginia DeJohn Anderson, *Creatures of Empire.*
44. Ritvo, "Emergence of Modern Petkeeping."
45. Darwin, *On the Origin of Species,* 7–43.
46. Clutton-Brock, *Natural History of Domesticated Mammals,* 196–97.
47. Rare Breeds Survival Trust website, http://www.rbst.org.uk/index.php.
48. Bailey and Culley, *General View,* 245.

— 13 —

Beasts in the Jungle (or Wherever)

When Byron wrote that "the Assyrian came down like the wolf on the fold" ("The Destruction of Sennacherib," 1815), his audience had no trouble understanding the simile or feeling its force, even though wolves had not threatened most British flocks since the Wars of the Roses. Almost two centuries later, expressions such as "the wolf is at the door" remain evocative, although the Anglophone experience of wolves has diminished still further. For most of us, they are only to be encountered (if at all) in zoos, or in establishments like Wolf Hollow, which is located in Ipswich, just north of Boston, where a pack of gray wolves lives a sheltered suburban existence behind a high chain-link fence. Their Massachusetts captivity has produced some modification of their nomadic habits and their fierce independent dispositions. (The pack was established twenty years ago with pups, so that only inherent inclinations needed to be modified, not confirmed behaviors.) Their relationship with their caretakers seems affectionate and playful, sometimes even engagingly doglike—so much so that visitors need to be warned that it would be very dangerous for strangers to presume on this superficial affability. The animals themselves give occasional indications that they retain the capacities of their free-roaming relatives—that though apparently reconciled to confinement, they are far from tame. When large loud vehicles rumble past on nearby Route 133, the wolves tend to howl. And despite their secure enclosure within the built-up landscape of North Ameri-

---

"Beasts in the Jungle (or Wherever)" originally appeared in *Daedalus* 137, no. 2 (Spring 2008): 22–30 (© 2008 by the American Academy of Arts and Sciences).

can sprawl, their calls evoke the eerie menace that has immemorially echoed through the wild woods of fairy tale and fable.

The symbolic resonance of large ferocious wild animals—the traditional representatives of what seems most threatening about the natural world—has thus proved much more durable than their physical presence. Indeed, their absence has often had equal and opposite figurative force. Thus the extermination of wolves in Great Britain, along with such other unruly creatures as bears and wild boars, was routinely adduced as evidence of the triumph of insular (as opposed to Continental) civilization in the early modern period. As they dispersed around the globe, British settlers and colonizers set themselves parallel physical and metaphorical challenges, conflating the elimination of dangerous animals with the imposition of political and military order. In North America, hunters could claim bounties for killing wolves from the seventeenth century into the twentieth, although by the latter period wolves had abandoned most of their historic range, persisting only in remote mountains, forests, and tundras. In Africa and (especially) Asia, imperial officials celebrated the "extermination of wild beasts" as one of "the undoubted advantages . . . derived from British rule."[1]

Very occasionally, large aggressive predators could symbolize help rather than hindrance. They served as totems for people whose own inclinations were conventionally wolfish or leonine. And alongside the legendary and historical accounts of big bad wolves existed a minority tradition that empha-

"The Wolf," from Thomas Bewick, *General History of Quadrupeds*, 1824.

sized cooperation rather than competition. From this perspective the similarities of wolf society to that of humans implicitly opened the possibility of individual exchange and adoption. A slender line of imagined lupine nurturers ran from the foster mother of Romulus and Remus to Akela, who protects and mentors Mowgli in *The Jungle Book* (1894). But in this way, as in others, Kipling's animal polity looked toward the past rather than the future. By the late nineteenth century, human opinions of wolves and their ilk had indeed become noticeably mixed. The cause of this amelioration, however, was not an altered understanding of lupine character or an increased appreciation of the possibilities of anthropo-lupine cooperation, but rather a revised estimation of the very qualities that had made wolves traditional objects of fear and loathing.

The shift in European aesthetic sensibility that transformed rugged mountains into objects of admiration rather than disgust is a commonplace of the history of aesthetics. For example, in the early eighteenth century, even the relatively modest heights of what was to become known as the English Lake District impressed Daniel Defoe as "eminent only for being the wildest, most barren and frightful of any that I have passed over in England, or even in Wales itself."[2] The increasingly Romantic tourists who followed him gradually learned to appreciate this harsh, dramatic landscape, so that a century later the noted literary opium eater Thomas DeQuincey could characterize the vistas that had horrified Defoe as a "paradise of virgin beauty."[3] Of course, this altered perception had complex roots, but it is suggestive that it coincided with improvements in transportation and other aspects of the infrastructure of tourism. As economic and technological developments made the world seem safer and more comfortable, it became possible to experience some of its extremes as thrilling rather than terrifying. Or, to put it another way, as nature began to seem a less overwhelming opponent, the valence of its traditional symbols began to change. Ultimately (much later, after their population numbers and geographic ranges had been radically reduced) even wild predators began to benefit from this reevaluation. The ferocity and danger associated with wolves and their figurative ilk became a source of glamour, evoking admiration and sympathy from a wide range of people who were unlikely ever to encounter them. As representatives of the unsettled landscapes in which they had managed to survive, they inspired nostalgia rather than antagonism.

Symbolic shifts were supplemented by shifts in scientific understanding, which redefined high-end predators as a necessary element of many natural ecosystems. Late-nineteenth-century attempts at wild animal protection were modeled on the hunting preserves of European and Asian elites. Thus the immediate antecedents of modern wildlife sanctuaries and national parks were designed to protect individual species that were identified as both desirable (whether intrinsically or as game) or in danger of extinction, whether the bison in North America or the giraffe in Africa. They were much less concerned with preserving the surrounding web of life. In most cases, indeed, early wildlife management policies had the opposite effect. Although not all of the species targeted for protection provided conventional hunting trophies—for example, by the end of the nineteenth century, many great ape populations were receiving some form of protection—all were herbivores. Further, none offered significant resistance to human domination of their territory. (If they did, policies could be reversed. For example, hippopotami, which enjoyed protection in some parts of southern Africa, were slaughtered with official encouragement in Uganda, where their belligerent attitude toward river traffic interfered with trade.[4]) Predators inclined to kill the species designated for protection received no protection themselves, either physical or legal. On the contrary, in many settings people simply replaced large predators at the top of the food chain and showed no mercy to their supplanted rivals.

Deep ancient roots can be unearthed for holistic or ecological thinking. Although most of the British pioneers of game preservation had enjoyed the classical education prescribed for privileged Victorian boys, the works of Charles Darwin may have offered more readily accessible arguments for understanding biological assemblages as interconnected wholes. Darwin provided many illustrations of the subtle and complex relationships among the organisms that shared a given territory. For example, in *On the Origin of Species,* he explained the frequency of several species of wildflowers in southern England as a function of the number of domestic cats kept in nearby villages. The cats had no direct interest in the flowers, but more cats meant fewer field mice, which preyed on beehives—therefore fewer mice meant more bees to fertilize the flowers.[5] Nevertheless, it was not until the second half of the twentieth century that individual species were routinely considered components of larger systems by wildlife managers, and that the standard unit of management became the ecosystem rather than the species. In consequence, large predators were redefined as essential components (even indicators) of a healthy environment rather than blots on the landscape. They often began to receive legal protection, however belated and ineffective. And there has been a movement to reintroduce them to areas that have been ostensibly preserved in their wild form or that are in process of restoration. Thus in recent decades wolves have reoccupied several of their former habitats in the western United States, both as a result of carefully coordinated reintroduction by humans, as in Yellowstone National Park, and as a result of independent (but unimpeded) migration from Canada. It is interesting that the actual or even the prospective reemergence of the wolf has inspired a parallel reemergence of traditional fear and hostility among neighboring human populations.

I have been using several terms as if their meanings were clear and definite, when in fact they are contested and ambiguous. As has often been repeated, the cultural critic Raymond Williams characterized "nature" as "perhaps the most complex word in the English language."[6] The term "wilderness" is similarly problematic. In the context of preservation or restoration, it often collocates with words like "pristine" and "untouched," and therefore connotes a condition at once primeval and static. This connotation suggests that the first task of landscape stewards is to identify this ur-condition, but even a moderately long chronological perspective suggests that any such effort is bound to be quixotic. The environment in which modern animals have evolved has never been stable. Less than twenty thousand years ago much of North America and Eurasia was covered by glaciers. After their gradual release from the burden of ice and water, most northern lands continued to experience significant shifts in topography and climate, and, therefore, in flora and fauna. These natural changes have been supplemented for thousands of years by the impact of human activities. The theoretical and political problems presented by "wilderness" are knottier still. In a groundbreaking essay published more than a decade ago, William Cronon argued that wilderness and civilization (or "garden") were not mutually exclusive opposites, but rather formed part of a single continuum. Far from being absolute, "the one place on earth that stands apart from humanity," wilderness was itself "a quite profoundly human creation."[7] Then as now, Cronon's formulation sparked agonized resistance on the part of environmentalists who based their commitment on the notion of untouched nature.

If wildness in landscape has been effectively (if controversially) problematized, the same cannot be said for wildness in animals. The *Oxford English Dictionary* defines the adjective "wild" unambiguously, and it emphasizes its zoological application. The first sense refers to animals: "Living in a state of nature; not tame, not domesticated: opp. to TAME." In a standard lexicographical ploy, "tame" is defined with equal confidence and complete circularity as (also the first sense) "Reclaimed from the wild state; brought under the control and care of man; domestic; domesticated. (Opp. to wild.)." But outside the dictionary these terms are harder to pin down and their interrelationships are more complex. Like Cronon's wilderness and garden, the wild and the tamed or domesticated exist along a continuum. In a world where human environmental influence extends to the highest latitudes and the deepest seas, few animal lives remain untouched by it. At least in this sense, therefore, few can be said to be completely wild—for example, it would be difficult so to characterize the wolves that were captured, sedated, airlifted to Yellowstone, and then kept in "acclimatization pens" to help them adapt to their new companions and surroundings. And as the valence of the wild has increased and its definition has become more obviously a matter of assertion rather than description, the boundaries of domestication have also blurred.

Not that they were ever especially clear. As twenty-first-century wolves belong to a long line of animals whose wildness has been compromised, tameness has also existed on a sliding scale. According to the *Oxford English Dictionary,* both "wild" and "tame" have persisted for a millennium, remaining constant in form as well as in core meaning, while the language around them has mutated beyond easy comprehension, if not beyond recognition. But this robustness on the level of abstraction has cloaked imprecision and ambiguity on the level of application or reference. Although medieval farmers and hunters may have had no trouble distinguishing livestock animals from game or vermin, it would have been difficult to extract any general definition from their practices. The impact of domestication varied from kind to kind, as well as from creature to creature. The innate aggression of the falcons and ferrets who assisted human hunters was merely channeled, not transformed; when they were not working, they were confined like wild animals in menageries. Then as now, people exerted much greater sway over their dogs than over their cats, who were mostly allowed to follow their own instincts with regard to rodents and reproduction. Medieval cattle, the providers of labor as well as meat, milk, and hides, led more constrained lives than did contemporary sheep, and pigs were often left to forage in the woods like the wild boars they closely resembled. With hindsight, even these relatively tame cattle might appear undomesticated, especially as wildness gained in glamour. Thus changes in the animals' physical circumstances were complicated by changes in the way they were perceived. By the late eighteenth century, for example, a few small herds of unruly white cattle, who roamed like deer through the parks of their wealthy owners, were celebrated as aboriginal and wild.

Only a few people possessed the resources necessary to express their admiration for the wild, and their somewhat paradoxical desire to encompass it within the domestic sphere, on such a grand scale. But numerous alternative options emerged for those with more restricted acres and purses. An increasing variety of exotic animals stocked private menageries. The largest of these were on a sufficiently grand scale to also include a cattle herd, if their owners had been so inclined—for example, those of George III and the thirteenth Earl of Derby accommodated large animals such as kangaroos, cheetahs, zebras, and antelopes. Smaller animals required more modest quarters, and parrots, monkeys, canaries, and even the celebrated but ill-fated wombats owned by the poet Dante Gabriel Rossetti could be treated as pets. Breeders attempted to enhance or invigorate their livestock with infusions of exotic blood. If they were disinclined or unable to maintain their own wild sire, they could, in the 1820s and 1830s, pay a stud fee to the newly established Zoological Society of London for the services of a zebu or a zebra. In Australia, Russia, Algeria, and the United States, as well as in Britain and France, the acclimatization societies of the late nineteenth century targeted an impressive range of species for transportation and domestication, from the predictable (exotic deer and wild sheep) to the more imaginative (yaks, camels, and tapirs).[8] So difficult (or undesirable) had it become to distinguish between wild animals and tame ones, that exotic breeds of domestic dogs were exhibited in Victorian zoos, and small wild felines were exhibited in some early cat shows.

The popular appeal of wild animals has continued to increase as they have become more accessible, either in the flesh or in the media. So entangled have wildness and domesticity become that it is now necessary to warn visitors to North American parks that roadside bears may bite the hands that feed them, and it is now possible for domesticated animals to represent nature. This extended symbolic reach was demonstrated in 2001, when foot-and-mouth disease struck British livestock. Although the outbreaks were widespread, the greatest number of cases occurred in the Lake District, and many of the threatened sheep belonged to the local Herdwick breed. Despite strong historical indications that the ancestral Herdwicks had arrived in the vicinity of the Lake District by boat, and the further fact that all British sheep descend

"A Mouflon Ram," from Richard Lydekker, *The Sheep and Its Cousins*, 1913.

from wild mouflons originally domesticated in the eastern Mediterranean region, they were traditionally celebrated as indigenous, "peculiar to that high, exposed, rocky, mountainous district."[9]

If vernacular usage illustrates the increasing slippage between wildness and tameness in animals, scientific classification has made a similar point from the opposite direction. The species concept has a long and vexed history. The study of natural history (or botany and zoology) requires that individual kinds be labeled, but in the case of many plants and animals (those that, unlike giraffes, for example, have very similar relatives) it has been difficult for naturalists to tell where one kind ends and the next begins. Darwin's theory of evolution by natural selection provided a theoretical reason for this difficulty, and his shrewd observations that "it is in the best-known countries that we find the greatest number of forms of doubtful value" and that "if any animal or plant . . . be highly useful to man . . . varieties of it will almost universally be found recorded" offered a more pragmatic explanation.[10] The classification of domesticated animals has epitomized this problem. That is, none of them has become sufficiently different from its wild ancestor to preclude the production of fertile offspring (the conventional if perennially problematic definition of the line between species), and some mate happily with more distant relatives. Nineteenth-century zookeepers enjoyed experimenting along these lines, and zoogoers admired the resulting hybrids between horses and zebras, domestic cattle and bison, and dogs and wolves.[11]

Despite these persuasive demonstrations of kinship, however, since the eighteenth-century emergence of modern taxonomy, classifiers have ordinarily allotted each type of domestic animal its own species name. While recognizing the theoretical difficulties thus produced, most modern taxonomists have continued to follow conventional practice. Domestic sheep are still classified as *Ovis aries* while the mouflon is *Ovis orientalis*, and dogs as *Canis familiaris* while the wolf is *Canis lupus*. The archaeozoologist Juliet Clutton-Brock explains this practice as efficient (it would be unnecessarily confusing to alter widely accepted nomenclature) as well as scientifically grounded, at least to some extent (most domestic animal populations are reproductively isolated from wild ones by human strictures, if not by biological ones).[12] But it also constitutes a simultaneous acknowledgment of the artificiality of the distinction between wild animals and domesticated ones, and of its importance and power. Vernacular understandings can trump those based on anatomy and physiology.

The implications of making or not making such distinctions extend beyond the intellectual realm. They both construct the physical world and describe it. Although the howls of the wolf may retain their primordial menace, the wolves who make them have long vanished from most of their vast original range, and they are threatened in much of their remaining territory. To persist or to return, they need human protection, not only physical but legal and taxonomic. With the advent of DNA analysis in recent decades, the taxonomic stakes have risen, so that even animals that look and act wild may be found genetically unworthy. Thus efforts to preserve the red wolf, which originally ranged across the southeastern states, have been complicated by suggestions that it is not a separate species, but a hybrid of the grey wolf and the coyote. Although no such aspersions have been cast upon the pedigree of the grey wolf, every attempted grey wolf restoration has triggered human resistance, and local challenges to their endangered status inevitably follow even moderate success. If domestic dogs were returned to their ancestral taxon, wolves would become one of the commonest animals in the lower forty-eight states, rather than one of the rarest. Their survival as wild animals depends on the dog's continuing definition as domesticated.

Notes

1. Lockwood, *Natural History, Sport, and Travel*, 237.
2. Defoe, *Tour Through the Whole Island of Great Britain*, 291.
3. DeQuincey, *Literary Reminiscences*, 311.
4. Ritvo, *Animal Estate*, 284–89.
5. Darwin, *On the Origin of Species*, 73–74.
6. Williams, *Keywords*, 184.
7. Cronon, "Trouble with Wilderness," 69.
8. Ritvo, *Animal Estate*, 232–42.
9. Bailey and Culley, *General View*, 245.
10. Darwin, *On the Origin of Species*, 50.
11. Ritvo, *Platypus and the Mermaid*, 92–95.
12. Clutton-Brock, *Natural History of Domesticated Mammals*, 194–97.

Bibliography

Periodicals

Agricultural Magazine, Plough, and Farmers' Journal
American Naturalist
Annals and Magazine of Natural History
Annals of Sporting and Fancy Gazette
Athenaeum
Banffshire Journal
Bazaar, the Exchange and Mart
Daily Graphic
Dog Owners' Annual
Dogs
Evening Dispatch
Express
Farmer and Naturalist
Field
Hippiatrist and Veterinary Journal
Illustrated London News
Independent
Irish Naturalist
Journal (Newcastle)
Journal of Agriculture
Land and Water
Land Magazine
Live Stock Journal
Magazine of Natural History
Manchester Guardian
Naturalist
Nature Notes
North Staffordshire Field Club Annual Report and Transactions
Notes and Queries
Park Cattle Society's Herd Book
Penny Magazine
Polo Magazine
Proceedings of the Royal Institution of Great Britain
Proceedings of the Zoological Society of London
Quarterly Journal of Agriculture
Quarterly Review
Quarterly Review of Agriculture
Science
Scotsman
Sketch
Spectator
Sportsman's Journal and Fancier's Guide
Times
Transactions of the Natural History Society of Glasgow
Transactions of the Tyneside Naturalists Field Club
Yorkshire Post
Zoologist

Other Sources

Alderson, G. L. H. "The History, Development, and Qualities of White Park Cattle." *Ark* 15 (1988): 126–28.

Alderson, Lawrence. "Foot-and-Mouth Disease in the United Kingdom 2001: Its Cause, Course, Control and Consequences." Paper presented at RBI/EAAP/FAO meeting, Budapest, 23 August 2001. http://www.warmwell.com/aldersonsept3.html.

Allen, David Ellison. *The Naturalist in Britain: A Social History.* Harmondsworth, Middlesex: Penguin Books, 1978.

Allen, Lewis Falley. *History of the Short-horn Cattle: Their Origin, Progress and Present Condition.* Buffalo, NY: Lewish Falley Allen, 1874.

Anderson, James. *Essays Relating to Agriculture and Rural Affairs.* Vol. 2. Edinburgh: William Creech, 1777.

Anderson, Virginia DeJohn. *Creatures of Empire: How Domestic Animals Transformed Early America.* New York: Oxford University Press, 2004.

"Animal Experiment and Medical Research: A Study in Evolution." *Conquest* 169 (February 1979): 1–14.

The Animal Museum; or, Picture Gallery of Quadrupeds. London: J. Harris, 1825.

"The Animal Rights Movement in the United States: Its Composition, Funding Sources, Goals, Strategies and Potential Impact on Research." Clarks Summit, PA: Society for Animal Rights, 1982.

Animal Sagacity, exemplified by facts showing the force of instinct in beasts, birds, &c. Dublin: W. Espy, 1824.

Bailey, John, and George Culley. *General View of the Agriculture of Northumberland, Cumberland, and Westmorland.* 1805. Reprint, Newcastle: Frank Graham, 1972.

Bajema, Carl Jay, ed. *Artificial Selection and the Development of Evolutionary Theory.* Stroudsburg, PA: Hutchinson Ross, 1982.

Banks, Joseph. *The Sheep and Wool Correspondence of Sir Joseph Banks, 1781–1820.* Edited by Harold B. Carter. London: British Museum (National History) for the Library Council of New South Wales, 1979.

Bates, Cadwallader John. *Thomas Bates and the Kirklevington Shorthorns: A Contribution to the History of Pure Durham Cattle.* Newcastle upon Tyne: Robert Redpath, 1897.

Bates, Thomas, and Thomas Bell. *The History of Improved Short-horn or Durham Cattle, and of the Kirklevington Herd, From the Notes of the Late Thomas Bates.* Newcastle: Robert Redpath, 1871.

Bennet, Ian. "Chillingham Cattle." *Ark* 18 (1991): 22.

Bennett, Edward Turner. *The Tower Menagerie: Comprising the Natural History of the Animals in that Establishment; with Anecdotes of their Characters and History.* London: Robert F. Jennings, 1829.

Bewick, Thomas. *A General History of Quadrupeds.* Newcastle upon Tyne: Beilby & Bewick, 1822.

Bibliography

———. *A General History of Quadrupeds*. Newcastle upon Tyne: Bewick and Son, 1824.

Bidwell, E., et al. "Report of the Committee . . . on the Herds of Wild Cattle in Chartley Park and Other Parks in Great Britain." *Report of the British Association for the Advancement of Science*, 1887, 135–45.

Bingley, Thomas. *Stories Illustrative of the Instincts of Animals, Their Characters and Habits*. London: Charles Tilt, 1840.

Bingley, William. *Animal Biography: or, Authentic Anecdotes of the Lives, Manners, and Economy of the Animal Creation, arranged according to the system of Linnaeus*. 3 vols. London: Richard Phillips, 1804.

Blacklock, Ambrose. *Treatise on Sheep; with the Best Means for their Improvement, General Management, and the Treatment of their Diseases*. Glasgow: W. R. McPhun, 1838.

Blumenbach, Johann Friedrich. *The Anthropological Treatises of Johann Friedrich Blumenbach*. Edited and translated by Thomas Bendyshe. London: Longman, Green, Longman, Roberts, & Green, 1865.

Boakes, Robert A. *From Darwin to Behaviourism: Psychology and the Minds of Animals*. Cambridge: Cambridge University Press, 1984.

Board of Agriculture. Minute book. 27 November 1798–18 March 1805. Institute of Agricultural History and Museum of English Rural Life, Reading, UK.

Bondeson, J., and A. E. W. Miles. "Julia Pastrana, the Nondescript: An Example of Congenital Generalized Hypertrichosis Terminalis with Gingival Hyperplasia." *American Journal of Medical Genetics* 47 (1993): 198–212.

Boreman, Thomas. *A Description of Three Hundred Animals, viz. Beasts, Birds, Fishes, Serpents, and Insects*. 3rd ed. London: R. Ware, 1736.

Bradley, A. G. *The Romance of Northumberland*. London: Methuen, [1908].

British Berkshire Herd Book. Vol. 1. Salisbury, Wilts: Edward Roe, 1885.

British Goat Society Herd Book and Prize Record from 1875 to 1885. Vol. 1. 1886.

Browne, Janet. *Charles Darwin: Voyaging*. New York: Knopf, 1995.

Buckley, Arabella. *The Winners in Life's Race, or the Great Backboned Family*. New York: D. Appleton, 1883.

Burkhardt, Richard W. "Closing the Door on Lord Morton's Mare: The Rise and Fall of Telegony." In *Studies in the History of Biology*, edited by William Coleman and Camille Limoges, 3:1–21. Baltimore: Johns Hopkins University Press, 1979.

Cambridge University Museum of Zoology. "Additions to the Museum." Vol. 1, 1867–1902.

———. "History Index." Vol. 3, 1892–1897.

Carroll, Lewis. *The Annotated Alice*. Edited by Martin Gardner. New York: Bramhall House, 1960.

Chambers, J. D., and G. E. Mingay. *The Agricultural Revolution, 1750–1880*. London: B. T. Batsford, 1966.

Churchill, Frederick B. "Sex and the Single Organism: Biological Theories of Sexuality in Mid-Nineteenth Century." In *Studies in the History of Biology*, edited by William

Coleman and Camille Limoges, 3:139–77. Baltimore: Johns Hopkins University Press, 1979.

Clarke, John H. *M. Pasteur and Hydrophobia: Dr. Lutaud's New Work*. London: Victoria Street Society United with the International Association for the Protection of Animals from Vivisection, n.d.

———. *The Pasteur Craze*. London: Victoria Street Society United with the International Association for the Protection of Animals from Vivisection, n.d. Originally published in *Zoophilist*, April 1886.

Clutton-Brock, Juliet. "British Cattle in the Eighteenth Century." *Ark* 9 (1982): 55–59.

———. "The Definition of a Breed." In *Archaeozoology: Proceedings of the IIIrd International Archaeozoological Conference*, 1:35–44. Szczecin, Poland: Agricultural Academy, 1979.

———. *A Natural History of Domesticated Animals*. London: British Museum (Natural History), 1987.

Coates, George. *The General Short-Horned Herd-Book Containing the Pedigrees of Short-Horned Bulls, Cows, &c. of the Imported Durham Breed*. Otley, W. Yorks.: W. Walker, 1822.

Coleman, John, ed. *The Cattle of Great Britain: Being a Series of Articles on the Various Breeds of Cattle of the United Kingdom, their Management, &c*. London: The Field, 1875.

Convery, Ian, et al. "Death in the Wrong Place? Emotional Geographies of the UK 2001 Foot and Mouth Disease Epidemic." *Journal of Rural Studies* 21 (2005): 99–109.

Cronon, William. *Changes in the Land: Indians, Colonists, and the Ecology of New England*. New York: Hill and Wang, 2003 (1983).

———. "The Trouble with Wilderness, or, Getting Back to the Wrong Nature." In *Uncommon Ground: Rethinking the Human Place in Nature*, edited by William Cronon, 69–90. New York: W. W. Norton, 1995.

Crosby, Alfred. *The Columbian Exchange: Biological and Cultural Consequences of 1492*. 1972. Reprint, New York: Praeger, 2003.

———. *Ecological Imperialism: The Biological Expansion of Europe, 900–1900*. 1993. Reprint, Cambridge: Cambridge University Press, 2004.

Culley, George. *Observations on Live Stock, Containing Hints for Choosing and Improving the Best Breeds of the Most Useful Kinds of Domestic Animals*. London: G. G. Robinson, 1786.

———. *Observations on Live Stock, Containing Hints for Choosing and Improving the Best Breeds of the Most Useful Kinds of Domestic Animals*. London: G. Wilkie & J. Robinson, 1807.

Dalziel, Hugh. *British Dogs: Their Varieties, History, Characteristics, Breeding, Management, and Exhibition*. London: "Bazaar," 1879–80.

———. *The Collie: As a Show Dog, Companion, and Worker*. Rev. J. Maxtee. London: L. Upcott Gill, 1904.

Darton, F. J. Harvey. *Children's Books in England: Five Centuries of Social Life.* Rev. 3rd ed. Edited by Brian Alderson. Cambridge: Cambridge University Press, 1982.

Darwin, Charles. *The Descent of Man.* 1871. Reprint, New York: Modern Library, 1950.

———. *The Life and Letters of Charles Darwin, Including an Autobiographical Chapter.* Edited by Francis Darwin. 3 vols. London: John Murray, 1888.

———. *On the Origin of Species by Means of Natural Selection, or the Preservation of Favoured Races in the Struggle for Life.* 1859. Facsimile reprint, edited by Ernst Mayr. Cambridge, MA: Harvard University Press, 1964.

———. *The Variation of Animals and Plants under Domestication.* 2 vols. 2nd ed., rev., 1883. Reprint, Baltimore: Johns Hopkins University Press, 1998. Originally published in 1868.

Davies, C. J. *The Kennel Handbook.* London: John Lane, 1905.

Defoe, Daniel. *A Tour Through the Whole Island of Great Britain.* Edited and abridged by P. N. Furbank, W. R. Owens, and A. J. Coulson. New Haven: Yale University Press, 1991. Originally published in 1724–26.

Delap, P. "Mammals." In *Natural History of the Lake District,* edited by G. A. K. Hervey and J. A. G. Barnes, 176–94. London: Frederick Warne, 1970.

DeQuincey, Thomas. *Literary Reminiscences from the Autobiography of an English Opium Eater.* In *The Works of Thomas De Quincey,* vol. 3. Boston: Houghton Mifflin, 1851.

Desmond, Adrian, and James Moore. *Darwin.* London: Michael Joseph, 1991.

Dickinson, William. "On the Farming of Cumberland." *Journal of the Royal Agricultural Society of England* 13 (1852): 207–300.

Dodds, Madeleine Hope, ed. *A History of Northumberland.* Vol. 14. Newcastle: A. Reid, Sons, 1935.

Dolan, Thomas M. *Pasteur and Rabies.* London: George Bell & Sons, 1890.

Edwards, Peter. *The Horse Trade of Tudor and Stuart England.* Cambridge: Cambridge University Press, 1988.

Elliot, T. J., trans. *A Medieval Bestiary.* Boston: David R. Godine, 1971.

Ewart, James Cossar. Cutting Book, 1896–97. James Cossar Ewart Papers. Special Collections, Edinburgh University Library. Gen 134.

———. "Experimental Contributions to the Theory of Heredity. A. Telegony." *Proceedings of the Royal Society of London* 65 (1899): 243–51.

———. *Guide to the Zebra Hybrids, Etc. on exhibition at the Royal Agricultural Society's Show, York, together with a description of Zebras, Hybrids, Telegony, Etc.* Edinburgh: T. and A. Constable, 1900.

———. *The Penycuik Experiments.* London: Adam and Charles Black, 1899.

The Fables of Aesop, and others, with Designs on Wood. 1818. Reprint, Newcastle: T. Bewick, 1823.

Fairholme, Edward G., and Wellesley Pain. *A Century of Work for Animals: The History of the R.S.P.C.A. (1824–1924).* New York: E. P. Dutton, 1924.

Farley, John. *Gametes and Spores: Ideas About Sexual Reproduction, 1750–1914.* Baltimore: Johns Hopkins University Press, 1982.

Fell, Clare. *Early Settlement in the Lake Counties.* Clapham, Yorks: Dalesman Books, 1972.
[Fenn, Eleanor Frere]. *The Rational Dame: Or, Hints Towards Supplying Prattle for Children.* 4th ed. London: John Marshall, [1790?].
Fleming, George. *Rabies and Hydrophobia: Their History, Nature, Causes, Symptoms, and Prevention.* London: Chapman & Hall, 1872.
Flower, William Henry. *The Horse: A Study in Natural History.* London: Kegan Paul, Trench, Trübner, 1891.
"Foot and Mouth: How the North Lived through a Nightmare." Supplement to *Journal* (Newcastle), 21 January 2002.
Franklin, Sarah. "Sheepwatching." *Anthropology Today* 17 (2001): 3–9.
Fream, W. "The York Meeting, 1900." *Journal of the Royal Agricultural Society of England*, 3rd ser., 11 (1900): 405–42.
Freeman, R. B. "Children's Natural History Books before Queen Victoria." *History of Education Society Bulletin* 17 (Spring 1976): 7–21.
———. "Children's Natural History Books before Queen Victoria: A Handlist of Texts." *History of Education Society Bulletin* 18 (Autumn 1976): 6–34.
French, J. O. "An Inquiry Respecting the True Nature of Instinct, and of the Mental Distinction between Brute Animals and Man." *Zoological Journal* 1 (1824): 1–32.
French, Richard D. *Antivivisection and Medical Science in Victorian Society.* Princeton, NJ: Princeton University Press, 1975.
The Galloway Herd Book, Containing Pedigrees of Pure-Bred Galloway Cattle. Vol. 1. Dumfries, Scotland: Galloway Cattle Society, 1878.
The General Stud Book, Containing Pedigrees of Race Horses. . . . Vol. 1. 3rd ed. London: James and Charles Weatherby, 1827. Originally published in 1791.
Gilpin, Sawrey. "On the character and expression of Animals." Bodleian Ms. Eng. misc.d.585 [folio 5–21]. Bodleian Library, Oxford University.
Godlovitch, Stanley, and Roslind Godlovitch, eds. *Animals, Men, and Morals.* London: Gollancz, 1971.
Goodacre, Francis Burges. *A Few Remarks on Hemerozoology; or, the Study of Domestic Animals.* London, 1875.
Gosse, Philip Henry. *Natural History: Mammalia.* London: Society for Promoting Christian Knowledge, 1848.
Graham, Caz, ed. *Foot and Mouth—Heart and Soul: A Collection of Personal Accounts of the Foot and Mouth Outbreak in Cumbria, 2001.* Carlisle: BBC Radio Cumbria, 2001.
Graham, Frederica. *Visits to the Zoological Gardens.* London: George Routledge & Sons, 1853.
Gripaios, Peter, et al. "The Economic Impact of Foot and Mouth Disease." South West Economy Centre, University of Plymouth, June 2001.
Guerrini, Anita. *Experimenting with Humans and Animals: From Galen to Animal Rights.* Baltimore: Johns Hopkins University Press, 2003.

Bibliography

The Guide to Service: The Cook. London: Charles Knight, 1842.

Hall, Stephen J. G. "Running Wild." *Ark* 16 (1989): 12–49.

———. "The White Herd of Chillingham." *Journal of the Royal Agricultural Society of England* 150 (1989): 112–19.

Hall, Stephen J. G., and Juliet Clutton-Brock. *Two Hundred Years of British Farm Livestock.* London: British Museum (Natural History), 1989.

Hanger, George. *Colonel George Hanger, to All Sportsmen, and Particularly to Farmers, and Game Keepers.* London: privately printed, 1814.

Harrison, Brian. "Animals and the State in Nineteenth-Century England." *English Historical Review* 88 (October 1973): 786–820.

Harting, James Edmund. *British Animals Extinct Within Historic Times, with Some Account of British Wild White Cattle.* London: Trübner, 1880.

Harwood, Dix. *Love for Animals and How it Developed in Great Britain.* New York: Columbia University, 1928.

Hassall, Charles. *General View of the Agriculture of the County of Carmarthen.* London: W. Smith, 1794.

Hawkins, Benjamin Waterhouse. *Comparative Anatomy as Applied to the Purposes of the Artist.* London: Winsor & Newton, 1883.

Haydon, Benjamin R. *Lectures on Painting and Design.* 2 vols. London: Longman, Brown, Green, & Longmans, 1844–46.

Hervey, G. A. K. "Flowering Plants, Ferns and Mosses." In *Natural History of the Lake District,* edited by G. A. K. Hervey and John A. G. Barnes, 37–45. London: Frederick Warne, 1970.

An Historical Miscellany of the Curiosities and Rarities in Nature and Art. 5 vols. London: Champante & Whitrow, 1794–1800.

Hole, N. H. "Rabies and Quarantine." *Nature,* 18 October 1969, 244–46.

Holloway, William, and John Branch. *The British Museum; or Elegant Repository of Natural History.* 2 vols. London: John Badcock, 1803.

Howard, Martin. *Victorian Grotesque: An Illustrated Excursion into Medical Curiosities, Freaks and Abnormalities, Principally of the Victorian Age.* London: Jupiter Books, 1997.

Hull, David L. *Darwin and His Critics: The Reception of Darwin's Theory of Evolution by the Scientific Community.* Chicago: University of Chicago Press, 1973.

In Memory of Consul. Manchester, n.d.

Inquiry Report: An Independent Public Inquiry into the Foot and Mouth Disease Epidemic that Occurred in Cumbria in 2001. Carlisle: Cumbria County Council, 2002.

Jacoby, Karl. *Crimes against Nature: Squatters, Poachers, Thieves, and the Hidden History of American Conservation.* Berkeley and Los Angeles: University of California Press, 2001.

James, Montague Rhodes. *The Bestiary, being a Reproduction in Full of the Manuscript Ii.4.26 in the University Library, Cambridge . . . and a Preliminary Study of the Latin Bestiary as Current in England.* Oxford: Roxburghe Club, 1928.

Jennings, John. *Domestic and Fancy Cats: A Practical Treatise on Their Varieties, Breeding, Management, and Diseases.* London: L. Upcott Gill, n.d.

Jesse, George Richard, comp. "Publications of Vivisection." [Scrapbook]. British Library.

———, comp. "Publications on Vivisection, 1875–1883." [Scrapbook]. British Library.

John Johnson Collection of Printed Ephemera. Bodleian Library, Oxford University.

Johnson, Edgar. *Sir Walter Scott: The Great Unknown.* 2 vols. New York: Macmillan, 1970.

[Jones, Stephen]. *The Natural History of Beasts, Compiled from the Best Authorities.* London: E. Newbery, 1793.

Kaplan, Colin. "The World Problem." In *Rabies: The Facts,* edited by Colin Kaplan, 7–8. Oxford: Oxford University Press, 1977.

Kean, Hilda. *Animal Rights: Political and Social Change in Britain since 1800.* London: Reaktion, 1998.

Kennel Club. The Kennel Club minute book. December 1, 1874 to April 21, 1884. Kennel Club Archives, London.

The Kent Dog-Owners and the New Muzzling Order, A Verbatim Report of the Proceedings at the Board of Agriculture on Wednesday, 22 January 1890. London: Dog Owners Protection Association, 1890.

Kerr, Robert. *The Animal Kingdom or Zoological System, of the Celebrated Sir Charles Linnaeus; Class I. Mammalia. . . .* London: John Murray, 1792.

Kingsford, Anna. *Pasteur: His Method and its Results.* London: North London Anti-Vivisection Society, 1886.

Kingsley, Charles. *Alton Locke, Tailor and Poet: An Autobiography.* Vol. 2. New York: Fred de Fau, 1899.

Knight, Charles. *The Pictorial Museum of Animated Nature.* 2 vols. London: C. Knight, 1844.

Knox, Robert. *Great Artists and Great Anatomists: A Biographical and Philosophical Study.* London: J. Van Voorst, 1852.

———. *The Races of Men: A Philosophical Enquiry into the Influence of Race over the Destinies of Nations.* London: Renshaw, 1862.

Kramnick, Isaac. "Children's Literature and Bourgeois Ideology: Observations on Culture and Industrial Capitalism in the Later Eighteenth Century." In *Culture and Politics from Puritanism to the Enlightenment,* edited by Perez Zagorin, 203–40. Berkeley and Los Angeles: University of California Press, 1980.

Lansbury, Coral. *The Old Brown Dog: Women, Workers, and Vivisection in Edwardian England.* Madison: University of Wisconsin Press, 1985.

Laqueur, Thomas. "Orgasm, Generation, and the Politics of Reproductive Biology." In *The Making of the Modern Body: Sexuality and Society in the Nineteenth Century,* edited by Catherine Gallagher and Thomas Laqueur, 24–35. Berkeley and Los Angeles: University of California Press, 1987.

Lawrence, John. *A General Treatise on Cattle, the Ox, the Sheep, and the Swine: Comprehending Their Breeding, Management, Improvement, and Diseases.* London: H. D. Symonds, 1805.

———. *A General Treatise on Cattle, the Ox, the Sheep, and the Swine.* London: Sherwood, Gilbert, & Piper, 1808.

Lawrence, William. *Lectures on Comparative Anatomy, Physiology, Zoology, and the Natural History of Man; delivered at the Royal College of Surgeons in the Years 1816, 1817, and 1818.* London: R. Carlile, 1823.

Le Brun, Charles. *Conference of Monsieur Le Brun, Chief Painter to the French King, Chancellor and Director of the Academy of Painting and Sculpture; upon Expression, General and Particular.* London: John Smith, Edward Cooper, & David Mortier, 1701.

Leclerc, Georges Louis, comte de Buffon. *Barr's Buffon: Buffon's Natural History. . . With Notes by the Translator.* Vol. 6. London: H. D. Symonds, 1797.

Lennie, Campbell. *Landseer: The Victorian Paragon.* London: Hamilton, 1976.

Lévi-Strauss, Claude. *Totemism.* Boston: Beacon, 1963.

Linnaeus, Carolus. *Systema Naturae: Regnum Animale.* 1758. Reprint, London: British Museum (Natural History), 1956.

Lledo, Pierre-Marie. *Histoire de la vache folle.* Paris: Presses Universitaires de France, 2001.

Lloyd Morgan, Conwy. *Animal Behavior.* London: Edward Arnold, 1900.

———. "Limits of Animal Intelligence." *Fortnightly Review* 54 (1893): 223–39.

Lockwood, Edward. *Natural History, Sport, and Travel.* London: W. H. Allen, 1878.

Low, David. *The Breeds of the Domestic Animals of the British Islands.* 2 vols. London: Longman, Orme, Brown, Green, & Longmans, 1842.

Lutwyche, Richard, et al. "Special: Foot and Mouth and Rare Breeds." *Ark* 29 (2001): 97–107.

Lydekker, Richard. *Guide to the Specimens of the Horse Family (Equidae) exhibited in the Department of Zoology, British Museum (Natural History).* London: British Museum (Natural History), 1907.

———. *A Hand-Book to the British Mammalia.* London: W. H. Allen, 1895.

———. *Hand-book to the Carnivora. Pt. I. Cats, Civets, and Mungooses.* London: Edward Lloyd, 1896.

Lytton, Judith Neville. *Toy Dogs and Their Ancestors, Including the History and Management of Toy Spaniels, Pekingese, Japanese and Pomeranians.* New York: D. Appleton, 1911.

Macpherson, H. A. *A Vertebrate Fauna of Lakeland, Including Cumberland and Westmorland with Lancashire North of the Sands.* Edinburgh: David Douglas, 1892.

Martin, W. C. L. "The Sheep." In *The Farmer's Library: Animal Economy,* vol. 2. London: Charles Knight, 1849.

Mayr, Ernst. *The Growth of Biological Thought: Diversity, Evolution, and Inheritance.* Cambridge, MA: Harvard University Press, 1982.

McCabe, J. Bertram. "The White Cattle of Cadzow." *Nature Notes* 8 (December 1897): 245.

M'Combie, William. *Cattle and Cattle-Breeders.* Edinburgh: William Blackwood, 1867.
Melville, Elinor G. K. *A Plague of Sheep: Environmental Consequences of the Conquest of Mexico.* 1994. Reprint, Cambridge: Cambridge University Press, 1997.
Millais, Everett. *The Theory and Practice of Rational Breeding.* London: Fanciers Gazette, 1889.
Mills, John. *A Treatise on Cattle.* London: J. Johnson, 1776.
Montagu, Jennifer. *The Expression of the Passions: The Origin and Influence of Charles LeBrun's Conférence sur l'expression générale et particulière.* New Haven, CT: Yale University Press, 1994.
The Natural History of Animals: Beasts, Birds, Fishes, and Insects. 1818. Reprint, Dublin: Smith & Son, 1822.
The Natural History of Domestic Animals: Containing an Account of their Habits and Instincts and of the Services They Render to Man. Dublin: J. Jones, 1821.
The Naturalist's Pocket Magazine: or, Compleat Cabinet of the Curiosities and Beauties of Nature. Vol. 7. London: Harrison, Cluse, [1800].
The Norfolk and Suffolk Red Polled Herd Book. Vol. 1. 1874.
Ormond, Richard. *Sir Edwin Landseer.* Philadelphia: Philadelphia Museum of Art, 1981.
Oxford Down Flock Book. Vol. 1. London: Oxford Down Sheep Breeders' Association, 1889.
Page, William, ed. *Victoria Country History of the Counties of England: Staffordshire.* Vol. 1. 1908. Reprint, London: Dawson's, 1958.
Park, Mungo. *Travels in the Interior of Africa.* Dublin: P. Hayes, 1825.
Passmore, John. "The Treatment of Animals." *Journal of the History of Ideas* 36 (1975): 195–218.
Patterson, Sylvia. "Eighteenth-Century Children's Literature in England: A Mirror of Its Culture." *Journal of Popular Culture* 13, no. 1 (1979): 38–43.
Pawson, H. Cecil. *Robert Bakewell, Pioneer Livestock Breeder.* London: Crosby Lockwood, 1957.
Pearsall, W. H., and Winifred Pennington. *The Lake District: A Landscape History.* London: Collins, 1973.
Peel, C. V. A. *The Zoological Gardens of Europe, Their Histories and Chief Features.* London: F. E. Robinson, 1903.
Pennant, Thomas. *History of Quadrupeds.* 3rd ed. 2 vols. London: B. & J. White, 1793.
Pickering, Samuel F. *John Locke and Children's Books in Eighteenth-Century England.* Knoxville: University of Tennessee Press, 1981.
Plumb, J. H. "The First Flourishing of Children's Books." In *Early Children's Books and Their Illustration,* edited by J. H. Plumb and Gerald Gottlieb, xvii–xxx. New York: Pierpont Morgan Library; Boston: David R. Godine, 1975.
Pratt, Samuel Jackson. *Pity's Gift: A Collection of Interesting Tales to Excite the Compassion of Youth for the Animal Creation.* 1798. Reprint, Philadelphia: J. Johnson, 1808.

A Pretty Book of Pictures for Little Masters and Misses, or Tommy Trip's History of Beasts and Birds. 15th ed. London: Edwin Pearson, 1867.

Provenzo, Eugene Francis. "Education and the Aesopic Tradition." PhD diss., Washington University in St. Louis, 1976.

Pusey, Philip. "On the Present State of the Science of Agriculture in England." *Journal of the Royal Agricultural Society of England* 1 (1840): 1–20.

Quayle, Eric. *The Collector's Book of Children's Books.* New York: Clarkson N. Potter, 1971.

———. *The Ruin of Sir Walter Scott.* London: Hart-Davis, 1968.

Regan, Tom. *The Case for Animal Rights.* Berkeley and Los Angeles: University of California Press, 1983.

Rennie, James. *Alphabet of Zoology, for the Use of Beginners.* London: Orr, 1833.

"Report of the Special Committee on the Society's Show System." *Journal of the Royal Agricultural Society of England,* 3rd ser., 11 (1900): 65–85.

Ritvo, Harriet. "The Animal Connection." In *Humans, Animals, and Machines: Boundaries and Projections,* edited by James Sheehan and Morton Sosna, 68–84. Berkeley and Los Angeles: University of California Press, 1991.

———. *The Animal Estate: The English and Other Creatures in the Victorian Age.* Cambridge, MA: Harvard University Press, 1987.

———. "The Emergence of Modern Petkeeping." *Anthrozoos,* Winter 1987. Reprinted in *Animals and People Sharing the World,* edited by Andrew Rowan, 13–32. Hanover, NH: University Press of New England, 1988.

———. *The Platypus and the Mermaid, and Other Figments of the Classifying Imagination.* Cambridge, MA: Harvard University Press, 1997.

———. "Pride and Pedigree: The Evolution of the Victorian Dog Fancy." *Victorian Studies* 29, no. 2 (1986): 227–53.

———. "The Roast Beef of Old England." In *Mad Cows and Modernity: Cross-disciplinary Reflections on the Crisis of Creutzfeldt-Jakob Disease,* edited by Iain McCalman, 97–123. Canberra: Australian National University, 1998.

Romanes, George. *Animal Intelligence.* New York: D. Appleton, 1896.

———. *The Life and Letters of George John Romanes.* Edited by Ethel Duncan Romanes. London: Longmans, Green, 1896.

———. *Mental Evolution in Animals.* London: Kegan, Paul, Trench, 1883.

Rowan, Andrew N., and Bernard E. Rollin, "Animal Research—For and Against: A Philosophical, Social, and Historical Perspective." *Perspectives in Biology and Medicine* 27, no. 1 (1983): 1–17.

Royal Society for the Prevention of Cruelty to Animals. *59th Annual Report.* 1885.

Rupke, Nicolaas A. *Vivisection in Historical Perspective.* London: Croom Helm, 1987.

Russell, Nicholas. *Like Engend'ring Like: Heredity and Breeding in Early Modern England.* Cambridge: Cambridge University Press, 1986.

———. "Who improved the eighteenth-century longhorn cow?" In *Agricultural*

Improvement: Medieval and Modern, edited by Walter Minchinton, 19–40. Devon: University of Exeter Press, 1981.
Ryder, M. L. *Sheep and Man*. London: Duckworth, 1983.
Salt, Henry S. *Animals' Rights Considered in Relation to Social Progress*. 1892. Reprint, Clarks Summit, PA: Society for Animal Rights, 1980.
Schwartz, Maxime. *How the Cows Turned Mad: Unlocking the Mysteries of Mad Cow Disease*. Translated by Edward Schneider. Berkeley and Los Angeles: University of California Press, 2003.
Scott, Sir Walter. *The Poetical Works of Sir Walter Scott*. Edited by J. Logie Robertson. London: Oxford University Press, 1904.
Sebright, John Saunders. *The Art of Improving the Breeds of Domestic Animals*. London: J. Harding, 1809.
Secord, James A. "'Nature's Fancy': Charles Darwin and the Breeding of Pigeons." *Isis* 72 (1981): 163–86.
Shaw, George. *General Zoology*. Vol. 1. London: G. Kearsley, 1800.
Shaw, Vero. *The Illustrated Book of the Dog*. London: Cassell, 1881.
Shoberl, Frederic. *Natural History of Quadrupeds*. 2 vols. London: John Harris, 1834.
Simmonds, Peter Lund. *The Curiosities of Food; or the Dainties and Delicacies of Different Nations Obtained from the Animal Kingdom*. London: R. Bentley, 1859.
Simmons, I. G. *The Moorlands of England and Wales: An Environmental History, 8000 BC–AD 2000*. Edinburgh: Edinburgh University Press, 2003.
Simpson, Frances. *Book of the Cat*. London: Cassell, 1903.
Sinclair, John. *The Code of Agriculture*. London: W. Bulmer, 1817.
Singer, Peter. *Animal Liberation: A New Ethics for Our Treatment of Animals*. New York: Avon Books, 1975.
Spargo, Demelza, ed. *This Land is Our Land: Aspects of Agriculture in English Art*. London: Royal Agricultural Society of England, 1989.
Stables, Gordon. *The Practical Kennel Guide; with Plain Instructions How to Rear and Breed Dogs for Pleasure, Show, and Profit*. London: Cassell Petter & Galpin, 1877.
Stephens, Frederick G. *Memoirs of Sir Edwin Landseer*. London: Bell, 1874.
Stevenson, Lloyd G. "Religious Elements in the Background of the British Anti-Vivisection Movement." *Yale Journal of Biology and Medicine* 19 (1956): 125–57.
Storer, John. *The Wild White Cattle of Great Britain: An Account of Their Origin, History, and Present State*. London: Cassell, Petter, Galpin, 1879.
Stubbs, George. *The Anatomical Works of George Stubbs*. Edited by Terence Doherty. London: Secker & Warburg, 1974.
Sussex Herd Book, Containing the Names of the Breeders, the Age, and the Pedigrees of the Sussex Cattle. Vol. 2. 1885.
Taplin, William. *The Sportsman's Cabinet, or, A Correct Delineation of the Various Dogs Used in the Sports of the Field*. Vol. 1. London, 1803.
Tavinor, Jan. "A Chapter in the History of the 'Chartleys.'" *Ark* 18 (1991): 378–80.
Teltruth, T. *The Natural History of Four-footed Beasts*. 3rd ed. London: E. Newbery, 1781.

Thomas, Keith. *Man and the Natural World: A History of the Modern Sensibility.* New York: Pantheon, 1983.

Thomas, Oldfield. *The History of the Collections Contained in the Natural History Departments of the British Museum.* Vol. 2. *Separate Accounts of the Several Collections Included in the Department of Zoology.* London: Trustees of the British Museum, 1906.

Thompson, Bruce. *The Lake District and the National Trust.* Kendal: Titus Wilson, 1946.

Thompson, Edward P. *The Passions of Animals.* London: Chapman & Hall, 1851.

Tom Trip's Museum: or, a Peep at the Quadruped Race. London: John Harris, n.d.

Topsell, Edward. *The Historie of Foure-Footed Beastes.* London: William Iaggard, 1607.

Trimmer, Mary. *A Natural History of the Most Remarkable Quadrupeds, Birds, Fishes, Serpents, Reptiles, and Insects.* Abr. ed. Boston: S. G. Goodrich, 1829.

Trimmer, Sarah Kirby. *Fabulous Histories: Designed for the Instruction of Children, Respecting Their Treatment of Animals.* 1786. Reprint, London: Whittingham & Arliss, 1815.

Trow-Smith, Robert. *A History of British Livestock Husbandry, 1700–1900.* London: Routledge & Kegan Paul, 1959.

Turner, James. *Reckoning with the Beast: Animals, Pain, and Humanity in the Victorian Mind.* Baltimore: Johns Hopkins University Press, 1980.

Tyson, Edward. *Orang-outang, sive Homo Sylvestris. Or, the Anatomy of a Pygmie Compared with that of a Monkey, an Ape, and a Man.* London: Thomas Bennet, 1699.

United Kingdom. Ministry of Agriculture and Fisheries. *British Breeds of Livestock.* London: Eyre & Spottiswoode, 1927.

United Kingdom. Parliament. *Report of a Committee appointed by the Local Government Board to inquire into M. Pasteur's Treatment of Hydrophobia.* c. 5087, LXVI.429, 1887.

———. *Report of the Departmental Committee to inquire into and report upon the working of the Laws relating to Dogs.* c. 8320, c. 8378, XXXIV, 1897.

———. *The Report from the Select Committee of the House of Lords on Rabies in Dogs.* (322) XI.451, 1887.

———. *The Select Committee on the Bill to Prevent the Spreading of Canine Madness.* 1830.

Varty, Thomas. *Graphic Illustrations of Animals, Showing Their Utility to Man, in Their Services During Life, and Use After Death.* London: privately printed, n.d.

Vasey, George. *A Monograph of the Genus Bos: The Natural History of Bulls, Bisons, and Buffaloes.* London: John Russell Smith, 1857.

Vernon, G. R., comp. *The Ayrshire Herd Book, Containing Pedigrees of Cows, Heifers, and Bulls of the Ayrshire Breed.* Vol. 1. Ayr, Scotland: Hugh Henry, 1878.

Walton, John. "Pedigree and the National Cattle Herd, circa 1750–1950." *Agricultural History Review* 34, pt. 2 (1986): 149–70.

Walton, John F. L. S., ed. *The Best Breeds of British Stock: A Practical Guide for Farmers and Owners of Live Stock in England and Colonies.* London: W. Thacker, 1898.

Walton, John K. "Mad Dogs and Englishmen: The Conflict over Rabies in Late Victorian England." *Journal of Social History* 13, no. 2 (1979): 219–39.

Walton, John R. "The Diffusion of Improved Sheep Breeds in Eighteenth- and

Nineteenth-Century Oxfordshire." *Journal of Historical Geography* 9 (1983): 173–95.
Watson, J. A. Scott. *The History of the Royal Agricultural Society of England, 1839–1939.* London: Royal Agricultural Society, 1939.
Webster, John. "Mammals in Cumbria—A Centenary Review." In *Cumbrian Wildlife in the Twentieth Century,* edited by David J. Clarke and Stephen M. Hewitt, 77–88. *Transactions of the Carlisle Natural History Society* 12. 1996.
Weir, Harrison. *Our Cats and All About Them.* Boston: Houghton Mifflin, 1889.
White, Charles. *An Account of the Regular Gradation in Man, and in Different Animals and Vegetables; and from the Former to the Latter.* London: C. Dilly, 1799.
Whitehead, G. Kenneth. *The Ancient White Cattle of Great Britain and Their Descendants.* London: Faber & Faber, 1953.
Wilkinson, John. *Remarks on the Improvement of Cattle, &c. in a Letter to Sir John Saunders Sebright, Bart.* Nottingham: H. Barnet, 1820.
Wilkinson, Lise. "The Development of the Virus Concept, as Reflected in Corpora of Studies on Individual Pathogens: 4. Rabies—Two Millennia of Ideas and Conjectures on the Aetiology of a Virus Disease." *Medical History* 21, no. 1 (1977): 15–31.
Williams, Raymond. *Keywords.* New York: Oxford University Press, 1976.
Wilson, Dudley. *Signs and Portents: Monstrous Births from the Middle Ages to the Enlightenment.* London: Routledge, 1993.
Wilson, Frank. *Westmorland Agriculture, 1800–1900.* Kendal: Titus Wilson, 1912.
Wilson, James. *The Evolution of British Cattle and the Fashioning of Breeds.* London: Vinton, 1909.
Winchester, Angus J. L. *The Harvest of the Hills: Rural Life in Northern England and the Scottish Borders, 1400–1700.* Edinburgh: Edinburgh University Press, 2000.
Wiseman, Julian. *The History of the British Pig.* London: Duckworth, 1986.
Wood, W. *Zoography; Or the Beauties of Nature Displayed.* London: Cadell & Davies, 1807.
Woods, Abigail. *A Manufactured Plague: The History of Foot and Mouth Disease in Britain.* London: Earthscan, 2004.
Wordsworth, William. *The Illustrated Wordsworth's Guide to the Lakes.* Edited by Peter Bicknell. New York: Congdon & Weed, 1984.
Youatt, William. *Cattle: Their Breeds, Management, and Diseases.* London: Baldwin & Craddock, 1834.
———. *Sheep: Their Breeds, Management, and Diseases.* London: Baldwin & Craddock, 1837.
Young, Arthur. *A Farmer's Tour Through the East of England.* 4 vols. London: W. Strahan, 1771.

Index

Italicized page numbers refer to illustrations.

Aesop's fables, 30
Africa, dangerous animals of, 43
agriculturalists, 116–17
AIDS, 61, 95, 97
Allom, Joseph, 159
Alton Locke (Kingsley), 9
American Association for the Advancement of Science, 59
American Medical Association, 59
Angola, monkeys of, 39
animal abuse, 35, 36; British legislation against, 75. *See also* antivivisectionism; vivisection
animal consciousness, 64; ideology and, 65
animal husbandry. *See* breeds and breeding
Animal Liberation (Singer), 52–53
Animal Museum, The, 36, 38–39
Animal Political Action Committee, 85
animal research. *See* antivivisectionism; vivisection
animal rights: animal protection laws in United Kingdom, 54; campaign for, 53. *See also* antivivisectionism
animals, 9; consciousness of, 67, 68, 70, 71; definition of, 3–4, 65; difference between humans and, 3, 9–10, 14, 38, 45, 65–66, 69–70, 86, 181; human-animal connection, 11, 182; as humans, 5–8, 30, 35, 38, 43; intelligence of, 68–69; in literature, 2, 30; mythical, 30, 33; qualities valued in, 68; treatment of, 36; utility of, 36, 37, 39–40. *See also* breeds and breeding; class, social; natural history; wild animals; *and specific breeds/species by name*
Animals (MSPCA magazine), 61
Animal Sagacity (anonymous), 38
animal studies, history of, 1–2, 11
Animal World, The, 58
Annals and Magazine of Natural History, 147
Annals of Sporting and Fancy Gazette, 114
antelope, 32, 42, 92
antivivisectionism: *Animal Liberation* and, 52–53, 74; animal rights and, 53; arguments over, 51; the Body Shop and, 51; campaigns against, 50–52, 61; commerce and, 51; complexity of issue, 59–62; continuity of ideas, 56, 59, 88; and Cruelty to Animals

Index

antivivisectionism *(continued)*
Act, 55, 76, 77; culture and, 53, 56, 62; decline of, 55–56, 83; dogs and, 77–78; as fad, 74; groups within movement, 55, 75, 77, 85, 88n5; history of, 50, 55–56, 74; human rights and, 56; *The Island of Dr. Moreau* and, 57–59; lasting solution, 61; law and, 50, 55, 61, 75; litigation and protest over, 51, 61; media and, 87; modern debate about, 59–61, 73, 83–88; MSPCA and, 50, 61; Pasteur and, 81–82; priorities of, 83; problems with, 61; public and, 76, 77, 80, 82, 83, 86–88; rabies and, 80–83; "regulationists" versus "abolitionists," 85; religious and philosophical objections to, 53–54, 74, 77, 82; RSPCA and, 75–76; scientific research, 56–57, 59, 86; SPCA and, 54; underestimation of, 74; in United Kingdom, 74–77, 82–83. *See also* humane movement; Public Responsibility in Medicine and Research (PRIM&R); rabies; vivisection

apes, 7, 8, 38, 65–66, 67, 178, 184; intelligence of, 69; Irish and, 178; in ranking of animals, 68, 69

Arctic Zoology (Pennant), 29

Ark (journal of Rare Breeds Survival Trust), 152

asses, 75, 172

Association for Biomedical Research, 85, 90n37

Attorneys for Animal Rights, 85

Atkins, Thomas, 114

aurochs, 144

Australia, 92

Babe (film), 10

baboon, 32; dog-faced, 35

Bacon, Francis, 53

badger, 42

Bakewell, Robert, 132, 153n1, 157, 191; background of, 157–58; breeding and, 132, 157, 160, 161–63, 166; and cattle, 157, 159, 160, 166, 167–68, 173; classification debate and, 173–74; controversy around, 158–61, 164; and Dishley sheep, 157, 159, 160, 161, 162, 173; and Dishley Society, 165; failures of, 160; fireside chair of, *159;* importance and influence of, 157–58, 161, 163, 173–74; inbreeding and, 160, 168, 175n10; naming of animals and, 168; non-Bakewellian breeding, 166, 167; pedigrees and, 167; rams of, 162; RASE and, 158; and seasonal hiring of animals, 163–66; success of, 162. *See also* breeds and breeding

bats, 69, 178

bears, 33, 36, 144, 184, 196; polar, 44

beavers, 9, 10, 33, 36, 144, 196

bestiaries, 30, 33, 43

Belle Vue Zoological Gardens (Manchester), 66

Bentham, Jeremy, 54, 59, 71, 75

Bewick, Thomas, 30, 33, 41, 139–40, 164, 172; *The Fables of Aesop,* 30; *A General History of Quadrupeds,* 33, 41, 139–40, 172; —, illustrations from, *16, 42, 143, 167, 204; Select Fables, with Cuts,* illustration from, *31*

birds, 38, 75, 109, 124

bison, 126, 206

Black Beauty (Sewell), 8, 46

Blair, Tony, 186

Blumenbach, Johann, 150

Board of Agriculture, 82, 161, 171

boars, 22, 144, 196, 208

Body Shop, 51

Boece, Hector, 146
Boethius, *Scoticorum Historiae,* 146
Booth, Thomas, 164
Boreman, Thomas, 29, 30, 31, 33–34, 47n10; *A Description of Three Hundred,* 29, 30, 33; —, illustration from, 32
Borlase, William, *Natural History of Cornwall,* 29
Boston Globe, 87
bovine spongiform encephalopathy (BSE): and affected cattle products, 97; animal-human transmission of, 92, 94–95; background of, 92; beef consumption and, 91, 92, 93, 94; British patriotism and mythology of, 98, 99, 100; cases in Europe, 95; cases in United Kingdom, 95; causes of, 95–96, 101; classification of, 92; foot-and-mouth disease outbreak and, 186, 188; initial British reaction to, 91–93, 94, 97, 101; McDonald's and, 92; nightmare scenario, 97; pattern of infection, 95; and policy making, 92, 93, 94–95, 98; political willpower and, 101; press and, 98; and preventative measures, 93, 94, 95–97, 101; and previous contamination alarms, 98; and pride in beef, 98–100; public relations and, 101; trade and, 91, 92, 93, 94, 100; transition of, 95–96; trivialization of, 92–94, 98, 101
Branch, John, *The British Museum; or Elegant Repository of Natural History,* 33, 43
breeds and breeding, 171; Bakewell cattle, 132, 167; Chillingham cattle, 132, 144, 145, 146; efficiency of, 124–25; females and, 13, 19, 20–24; Herdwick sheep and, 191–92; heredity and, 109, 110; and human gender, 17, 23–24, 26–27; hybrids and, 107–8, 113, 115–16; ignorance of science and, 116–18; illegal, 163–64; importance of, 24; limits of, 16, 126; males and, 17, 20, 22–23; naturalists versus agriculturalists, 116–18; public, 24, 25; scientific basis of, 17, 20, 27n10, 110, 113, 116, 118–19, 161; social aspects of, 14, 19, 24, 25, 117; status and, 171; stud books, 6, *18,* 99, 168–70, 176n43; Victorian "advances" in, 15–16, 17, 27n10; vocabulary of, 26, 27; women and, 14, 17, 23–24, 25, 26, 27. *See also* Bakewell, Robert
British Association for the Advancement of Science (BAAS), 103, 114, 139, 141, 142, 143
British Berkshire Pig Society, 169
British Food and Farming Year (1989), 158
British Goat Society, 169, 170
British Museum (Natural History), 139, 143, 184
British Museum; or Elegant Repository of Natural History, The (Holloway and Branch), 33, 43
Buckley, Arabella, 34–35, 44, 45, 46; *The Fairyland of Science,* 45; *The Winners in Life's Race,* 35, 44
bulls, 13, 19, 22, 39, *152, 170;* human identification with, 134; Royal, 104
Byron, Lord, 203

cabbage, 127
Cadyow Castle (Scott), 133–35, 137, 139, 148
Cadzow Castle, white cattle at, 132, 133, 135, 138, 139, 144, 145–46. *See also* Chillingham cattle
Cambridge University Museum of Zoology, 139
camels, 36, 42

Canada, wolves in, 207
Caplan, Arthur, 86
capybaras, 42
Carli, Father, 39
carnivores, 7; dangers of, 43
cattle, 15, 16, 37, 39, 41, 103, 118, 126, 146, 166, 173, 191, 208, 211; Ayrshires, 146; Bakewell and, 157, 159, 160; Carmarthenshire, 171; Castlemartin, 146; foot-and-mouth disease and, 187; Highland, 104; Irish moiled, 189; Kyloes, 146; at RASE shows, 104; Scottish wild, 135; *Variation of Animals and Plants under Domestication* and, 126. *See also* bovine spongiform encephalopathy (BSE); bulls; Cadzow Castle; Chillingham cattle
cats, 9, 10, 33, 38, 75, 206; antivivisectionism and, 51, 85; breeding of, 16, 19–20, 26; BSE in, 92, 93; hybrids, 105; Lake District wild cat, 196; problems with, 41; rabies and, 89n26; *Variation of Animals and Plants under Domestication* and, 127. *See also* wolves
Celts, 177
Chartley Park (Staffordshire), 138, 139, 142
Charles, William, "John Bull and the Alexandrians," 100
Cheshire, 137
chicken, 128
Chillingham, 132; Castle, 136
Chillingham cattle, *137, 138, 143, 152,* 172–73, 209; Ardrossan herd, 148; appeal of, 132; breeding of, 132, 144, 145, 146, 149; Britishness of, 133, 142, 144, 146, 148–49; Cadzow herd, 132, 133, 135, 138, 139, 144, 145–46; categorization of, 151, 153; Chillingham herd, 132, 136, 137, 140, 142, 145, 146, 147, 148,

150, 152–53, 153n7; descriptions of, 140–41, 156n62; as domesticated animals, 142–44, 152; gender and, 149, 150; in *A General History of Quadrupeds,* 139–40; horned, 142; human identification with, 137, 148–50; human racial purity and, 150–51; Kirkcudbrightshire herd, 147; Landseer and, 135–37, 141; location of herds, 132, 145, 146; Lyme herd, 137, 145, 147; in museums, 139; origins of, 142–45, 146, 147–48, 152, 153, 156nn68–69; popularity of, 132–33, 137–39, 151; Prince of Wales and, 137; problems studying, 148; protection of, 144; purity of herd, 141–42, 145, 150–51, 152; symbolism of, 133; as tabula rasa, 148–49; taxonomical status of, 139; threats to, 144; as tourist attraction, 138; as wild animals, 142, 153; writing on, 146–48. *See also* Tankerville, 5th Earl of; Tankerville, 6th Earl of; Tankerville, 7th Earl of
Chillingham Wild Cattle Association, 152
chimpanzees, 7, 11, 61, 65, 66; Consul, 66, *67;* as humans, 66, 179; Jenny, 66
class, social: animal hierarchy and human society, 37–38, 44, 46; antivivisectionism and, 56; carnivores and, 43; domesticated animals and, 41; human equality and, 65; middle-class education and natural history, 34, 44, 47n17; quadrupeds and, 39; Wordsworth and, 195
classification, 33, 47n8, 65, 171–74, 176n46, 181, 183–85, 210; experts in, 70–71; Herdwick sheep and, 191; human-animal, 65, 181, 183; problems with Linnaean system, 33, 69, 178–79. *See also* humans
Clutton-Brock, Juliet, 211

Coalition to Abolish LD-50 and Draize Test, 85
Colam, John, 76
Comparative Anatomical Exposition of the Human Body with that of a Tiger and a Common Fowl, A (Stubbs), 183–84; illustration from, *184*
Conservative Party (UK), 93, 94, 98, 99, 101
Consul (chimpanzee), 66, 67
cows, 13, 19, 20, 22, 40, 75. *See also* bovine spongiform encephalopathy (BSE); cattle
Creutzfeldt-Jakob Disease (CJD), 92, 95
Critical Period in the Development of the Horse, A (Ewart), 107
Cronon, William, 207, 208
Cruelty to Animals Act (1876), 55, 76, 77
Culley, George, 163, 173; *Observations on Livestock,* 140
Cumbria, 187, 192, 196. *See also* Lake District
Cuvier, George, 150

Dalziel, Hugh, 20
Darwin, Charles, 68, 69, 108, 181, 182, 200, 210; breeding and, 15, 17, 27n9, 34, 37, 116, 117, 124–25, 128, 129; Chillingham cattle and, 147, 148; modern view of, 130–31; publishing plans of, 123–24; as Victorian, 130–31; vivisection and, 55, 57, 84; A. R. Wallace and, 124; and wildlife preservation, 206; works of, 124, 130. *See also specific works by title*
Dawkins, W. Boyd, 143, 145
Deadly Feasts (Rhodes), 97
Death of the Wild Bull and Scene in Chillingham Park, The (Landseer), 135–36, 141, 153n7
deer, 42, 136

Defoe, Daniel, 205
Department of Agriculture (U.S.), 101
Derby, Earl of, 114, 209
Descartes, René, 68, 70, 71
Descent of Man, and Selections in Relation to Sex, The (Darwin), 70, 124, 182
Description of Three Hundred Animals, A (Boreman), 29, 30, 33; illustration from, *32*
DeQuincey, Thomas, 195, 196, 205
Dickens, Charles, *Household Words,* 99
Dishley sheep. *See under* Bakewell, Robert; sheep
Dishley Society, 165, 191
Dog Owners Protection Association, 82
dogs, 10, 13, 33, 41, 71, 75, 77, 144, 172, 211; antivivisectionism and, 51, 85; basset hounds, 116, border collies, 197–98; breeding of, 15, 20–21, 22, 23, 26, 38, 116, 161; deerhound, 135; English sheepdog, *198;* foxhound, 13, 161; greyhound, 161; hybrids, 114, 117, 180; intelligence and loyalty of, 40–41, 69, 182; rabies and, 77–80, 82, 88–89n26; in ranking of animals, 68; *Variation of Animals and Plants under Domestication* and, 126; vivisection and, 77–78, 87
Dolan, Thomas, 81
domesticated animals: breeding and, 171; classification of, 171–72, 200; connection to humans, 182; difference between domestic and wild animals, 200, 208–11; domestication of wild animals, 41–42, 209; *Origin of Species* and, 124; ranking of, 39–41; *Variation of Animals and Plants under Domestication* and, 126–27, 131. *See also* Chillingham cattle
domesticated plants, 127, 131
donkeys, 22, 105, 129

Edinburgh: Cattle Mart, 106; University of, 106, 109; Zoo, 109
Edward VII, 104, 106, 137; head of a Chillingham bull shot by, *138*
Egypt, 91
elephants, 33, 38; African, 9; domestication and, 41–42; Jumbo, 9
endangered species, 189. *See also* Rare Breeds Survival Trust
environment, perpetual changes in, 207–8. *See also* landscape; wildlife
European Union, 91, 93–94, 100
Exeter Change Menagerie, 66
Expression of the Emotions in Man and Animals, The (Darwin), 124, 131
Ewart, James Cossar, *113;* background and accomplishments of, 106, 109; breeding experiments and, 109, 112; *Guide to the Zebra Hybrids,* 106, 114, 115, 120; ––, illustration from, *113;* hybrid experiments and, 106, 108–12, 120; interest in birds, 109; Lord Morton's Mare and, 110–11; Penicuik farm of, 106, 107, 108, 110; *The Penycuik Experiments,* 107–8; publications of, 106, 107–8, 113, 114, 115, 120; publicity and, 106, 107–8, 113–14; research and breeding community, 117–18, 120; Royal Society presentation, 113; supermule experiments, 112; universal appeal of research, 108, 109, 114, 115, 118; vivisection and, 108–9. *See also* Matopo (zebra); quagga; Romulus (hybrid zebra); zebras
Eyton, Thomas, 116

Fables of Aesop, The (Bewick), 30
Fabulous Histories (Trimmer), 38
Fairyland of Science, The (Buckley), 45
"Farming Today" (BBC), 91

Farne Island, 138
Farrier and Naturalist, 13, 19
females, antivivisectionism and, 56. *See also* breeds and breeding; class, social; humans
Fenn, Eleanor Frere, *The Rational Dame,* 35, 36
Ferrers, Earl of, 138, 141
Fishery Board of Scotland, 109
Fleming, George, 80
foot-and-mouth disease: affected areas, 187; ecological impact of, 194; economic consequences of, 187; Lake District tourism and, 187–88; losses due to, 187; outbreak of, 186–87, 209; press and, 188; reaction to, 186–88, 199, 201n1. *See also* Herdwick flock; Lake District
Foot and Mouth—Heart and Soul, 187
foxes, 89–90n26
Fox Terrier Chronicle, 26
France, 99, 100
Fream, W., 114
Fry, John, 117

Galloway Cattle Society, 169
Garrard, George, "Woburn Sheepshearing," *165*
gender. *See* breeds and breeding; class, social; females, antivivisectionism and; humans
General History of Quadrupeds, A (Bewick), 33, 35, 41, 139–40; animal classification in, 172; illustrations from, *16, 42, 167, 204*
George III, 209
Germany, 112
Gesner, Konrad von, *Historia Animalium,* 30
Gilpin, Sawrey, 183
giraffes, 42, 210

Glasgow and West of Scotland Technical School, 107
goats, 22, 36, 41, 75
Goodacre, Frances, 143
Goodall, Jane, 61
Goldsmith, Oliver, 33; *An History of Earth and Animated Nature*, 29–30
gorillas, 11; Koko, 10, 12n11
Gosse, Phillip, 137
Grange in Borrowdale: Early Morning (Lake District scene), *194*
Graphic Illustrations of Animals (Varty), 36
Guide to the Lakes (Wordsworth), 195
Guide to the Zebra Hybrids (Ewart), 106, 114, 115, 120; illustration from, *113*
Gummer, John, 93

Hamilton, Lady Anne, 135
Hamilton Place, 135
Harting, James, 140
Harvard University, 85, 86
Hastings Institute, 86
Haydon, Benjamin, 183
Herdwick flock, 189, *190*, 197; breeding and, 191–92; classification of, 191; description of, 190–91, 193; foot-and-mouth disease and, 193–94, 198, 200, 209; intrinsic value of, 189, 197, 200; larger importance of, 198; location and numbers of, 189; National Trust and, 190; origins of, 192–93, 200, 209–10; press coverage of, 189; prizes for, 191–92; purity of breed, 192; RASE and, 191–92
Herdwick Sheep Breeders' Association, 190
heredity. *See* breeds and breeding; hybrids
Heritage Lottery Fund, 196
Highland Agricultural Show, 106
Hippiatrist and Veterinary Journal, 117
hippopotamus, 9, 10, 42, 206

Historia Animalium (Gesner), 30
Historie of Foure-Footed Beastes (Topsell), 30
history, 63, 64–65; development of, 131; environmental, 199; generalizations and, 69; potted history, 64, 71. *See also* natural history
History of Earth and Animated Nature, An (Goldsmith), 29–30
History of Quadrupeds (Pennant), 140; illustration from, *35*
HMS *Beagle*, 123
hogs, 33, 41
Holloway, William, *The British Museum; or Elegant Repository of Natural History*, 33, 43
Horse, The: A Study in Natural History, 184
horses, 16, 33, 36, 37, 41, 75, 103, 118, 126, 172, 211; Arabian, 105, 112, 117, 128; Bakewell and, 160; breeding, 22, 161, 167; carthorses, 42, 157; Cleveland bay, 189; Connemara pony, 109; draft, 15; Exmoor pony, 112; Highland pony, 112; Iceland pony, 111, 112; in Ireland, 109; Irish pony, 112; New Forest pony, 112; "nobility" of, 40; Norwegian pony, 112; racehorses, 15, 16, 161; Shetland pony, 105, 111, 112; Welsh pony, 112; West Highland pony, 112, *113*. *See also* Ewart, James Cossar; hybrids; Lord Morton's Mare
Horsley, Victor, 80
Household Words (Dickens), 99
humane movement, 75, 77
humans: as animals, 3, 5–8, 35, 46, 65–66, 67, 69–70, 86, 177–79, 181, 183–85, 204; Enlightenment naturalist view of, 65; equality of, 65; identification with animals, 134, 137; as separate from animals, 181–82; skeleton, *184*. *See also* breeds and breeding; class, social

Huxley, Thomas, 84
hybrids, 12n5, 179–80; cat, 105; cattle, 211; dog, 114, 117; ferret-stoat, 115; heredity and, 109–10; horse, 105, 180; lion-tiger, 114, *119;* "mental-hybridization," 181; mule, 112, 114; pigeon, 105; popularity of, 104, 106, 114; rabbit, 105; research applicability of, 115, 116–17; scientific information on, 106; stoat-ferret, 115; zebra, 104–5, 106, 111–14. *See also* Ewart, James Cossar; Romulus (hybrid zebra)
hyenas, *32,* 44

Illustrated London News, 104
Independent, 193–94
India, 112
insects, 9, 108–9, 112
Institute of Health Research, 188
Introduction to the Principles of Morals and Legislation, An (Bentham), 53–54
Irish, the, 177–78
Irish Naturalist, 108
Island of Dr. Moreau, The (Wells), 57–59
Issues in Science and Technology, 59

jackals, 44
Jenny (chimpanzee), 66
Jesse, George Richard, 77
John Bull, 99, *100*
Jones, Stephen, *Natural History of Beasts,* 34, 40, 43
Journal of the Royal Agricultural Society of England, 114, 118, 120
Jumbo (elephant), 9
Jungle Book, The (Kipling), 205

Kagan, Connie, 85
Kennel Club, 82, 163
Kent, Duchess of, 75

Kingsford, Anna, 77
Kingsley, Charles, *Alton Locke,* 9
Kipling, Rudyard, 205; *The Jungle Book,* 205
Kirkcudbrightshire, 147
Knavesmire, 103
Knox, Robert, 150, 156n68
kuru, 92

Labour Party (UK), 94, 101, 186
Ladies' Kennel Journal, 26
Lake District, *194;* animals of, 195, 196; ecological past of, 195, 197, 198; flora and fauna of, 197; foot-and-mouth disease outbreak in, 187–88, 209; human impact on, 195; importance of, 194; Natural Trust and, 190; sheep in, 189, 191; tourism, 187–88, 205; wildness of, 205. *See also* Herdwick flock
Lancaster University, 188
Lancet, 92, 108, 181
landscape, 207; in Europe, 199; history and, 199; in North America, 199
Landseer, Edwin, 135, 149; *The Death of the Wild Bull and Scene in Chillingham Park,* 135–36, 141, 153n7; Royal Academy and, 135, 137; *The Wild Cattle of Chillingham, 136,* 136–37, 141
Lawrence, John, 173
Lawrence, William, 181
LeBrun, Charles, 182–83
leporides, 180, *180*
Linnaean categories, 33, 69, 178–79
Linnaean Society, 124, 140
Linnaeus, Carl, 69, 172, 178; *Systema Naturae,* 179
lions, 33, 41, 133; honor of, 43; hybrid lion-tiger, 114, *119;* people as, 204
livestock, breeding of, 15, 19, 26. *See also* breeds and breeding

Locke, John, 34
Loeb, Jerod, 59
Loew, Franklin M., 84–85, 86
Lord Morton's Mare, 110, 111, 112, 115, 117, 128
Low, David, 146, 164; illustration from *Breeds of the Domesticated Animals*, 170
Lydekker, Richard, 143
Lyme Park (Cheshire), 137, 145, 147
lynx, 36
Lytton, Judith Neville, 13, 26

MacPherson, H. A., 195
MacPherson, James, 133
mad cow disease. *See* bovine spongiform encephalopathy (BSE)
Magazine of Natural History, 117
Maidstone, 103, 104
males. *See* breeds and breeding; class, social; females, antivivisectionism and; humans
mammals, study of, 34
Manchester Guardian, 104, 106, 108
Martin, Richard, 75
Massachusetts Institute of Technology (MIT), 57–58
Massachusetts Society for the Prevention of Cruelty to Animals (MSPCA), 50, 51, 61, 85
maucaucos, 69
Matopo (zebra), 104–5, 105, 107, 108, 111–12, 115
Mayr, Ernst, 70
McDonald's, 91
menageries, 66, 114, 139
Menageries, The (Rennie), illustration from, 45
Millais, Everett, 116, 118, 119
Ministry of Agriculture, Fisheries, and Food (MAFF), 94, 191
minks, 36

Minstrelsy of the Scottish Border, The (Scott), 133–34
monkeys, 7, 38, 39, 65, 69, 178, 184; purple-faced, 35
Monograph of the Genus Bos (Vasey), 141, 149
Morgan, Conwy Lloyd, 71
Morice, Humphrey, 75
Morning Post, 108
mules, 112, 114, 115
Musée d'Histoire Naturelle (Paris), 150
mythical animals, Scottish white cattle as, 135. *See also under* animals

National Trust, 190
National Veterinary Association of Ireland, 114
natural history, 46n3, 71; educational uses of, 29, 30, 34; format and content of books, 33–34, 36; history of, 30; large animals and, 43; literature for children, 29, 31–34, 35, 36, 37, 43, 44, 45, 46, 46n1, 47n17; medieval roots of, 30, 33, 43; morality and, 35–36, 37, 38, 42, 44–45, 46; natural order and, 44; neutrality on wild animals, 41–42; popularity of, 29, 53; practitioners of, 71; religious connections and, 34–35, 36; sources, 30, 33, 45, 71; and treatment of carnivores, 43. *See also* class, social
Natural History of Animals, The (anonymous), 42
Natural History of Beasts, The (Jones), 34, 40, 43
Natural History of Cornwall, The (Borlase), 29
Natural History of Four-footed Beasts, The (anonymous), 31, 33, 39
naturalists, 67, 68, 116–17, 128, 131, 171, 178, 181; continuity and, 70

Naturalist's Pocket Magazine, 172
Natural Science, 108
Newbery (publisher), 31, 33
Newcastle, 139
New Leicester sheep. *See* Bakewell, Robert; sheep
New Scientist, 98
Newton, Isaac, 70
New York Times, 87
New Zealand, 91
Northumberland, 132
Notes and Queries, 114
NOVA (PBS), 53

Observations on Livestock (Culley), 140
Oldham, Charles, 147
On the Origin of Species (Darwin), 69, 123–24, 177, 182, 200, 206; Darwin's concerns over, 123; evidence in, 124–25; pigeons in, 124–25; and superiority of natural selection, 125–26; weakness of, 129
opossum, 33
Orang-outang (Tyson), 65, 178, 179
orangutans, 4, 7, 38, 65, 66, 66, 179
"Origins and Early Progress of Our Breeds of Polled Cattle" (anonymous), 145
Ossulston, Lord (6th Earl of Tankerville). *See* Tankerville, 5th Earl of; Tankerville, 6th Earl of
Owen, Richard, 143, 144
Owens College, 143
ox, 13, 20, 33, 39, 41; aurochs, 144; Durham, 167
Oxford English Dictionary, 167, 171, 192, 207, 208

Park, Mungo, 43
Park Cattle Society, 153
Park Cattle Society's Herd Book, 148

Park Sheep Society, 109
Pasteur, Louis, 80–82
Pasteur Institutes, 81
Paton, W. D. M., 84
peccaries, 33
pedigree, breeding, 13, 14; history of, 167
Penicuik, 106, 107, 108, 110, 114, 115, 118
Pennant, Thomas, 69, 140, 172–73; *Arctic Zoology,* 29
Penrith Agricultural Society, 191
Penycuik Experiments, The (Ewart), 107–8
People for the Ethical Treatment of Animals (PETA), 51–52
Percy, Thomas, 133
pets, 24, 199–200
pigeons, 27n9, 105, 124, 127; English Fantail, 127
pigs, 10, 14, 15, 22, 75, 103, 104, 118, 133, 169, 173, 191, 208; Bakewell and, 157, 160; foot-and-mouth disease and, 187; hybrid, 114; uncooperativeness of, 41
Polo Magazine, 107
Potter, Beatrix, 197
Pratt, Samuel, 35
Pretty Book of Pictures for Little Masters and Misses, or Tommy Trip's History of Beasts and Birds, A (anonymous), 31–32, 33
Prichard, J. C. 150
Privy Council, 75
Proceedings of the Zoological Society of London, 114
Prusiner, Stanley, 96
Public Responsibility in Medicine and Research (PRIM&R), 73, 84–88
Punch, 183; cartoons, 25, 52, 54, 79

quadrupeds, 33, 34, 37, 38, 40, 46, 65, 171; inferiority of, 39

quagga, 110, *111*, 111, 115, 117, 128, 129
Quarterly Review, 108

rabbits, 60, 105
rabies, 77–78, 80, 82–83, 89–90n26; deaths from, 78, 82; government action and, 78–79, 80, 82, 89n20, 89n26; objections to cure, 81–83; and quarantine of United Kingdom, 80; scientific knowledge of, 78–79, 80, 81–82. *See also* Pasteur, Louis
Rabies and Hydrophobia (Fleming), 80
raccoon, 42
Rare Breeds Survival Trust, 152, 154n30, 189, 200
rats, 60
Rational Dame, The (Fenn), 35, 36
Reader's Guide to Periodical Literature, The, 87
Regent's Park Zoo, 66, 139, 209
Reliques of English Poetry (Percy), 133
rhinoceros, 42
Rhodes, Richard, 97
Romanes, George John, 68, 71, 109, 182
Romulus (hybrid zebra), 107
Rossetti, Dante Gabriel, 209
Rowan, Andrew, 86
Royal Academy, 135, 136
Royal Agricultural Society of England (RASE), 99, 103–5, 119; Bakewell and, 158; Herdwick sheep and, 191–92, hybrids and, 104–5, 106, 108, 115, 118
Royal Commission on the Practice of Subjecting Live Animals to Experiments for Scientific Purposes, 76
Royal Institution, 107
Royal Society, 106, 108, 113, 118
Royal Society for the Prevention of Cruelty to Animals (RSPCA): activities of, 77, 89n10; antivivisectionism and, 75–76; founding of, 75; rabies and, 82; scientific sympathies of, 75–76
Royal Zoological Society of Scotland, 109
Runkle, Deborah, 59

sagacity: in animals, 68–69; in humans, 68
Sainsbury's, 93
St. Albans, 147
Salt, Henry, 77, 88n6
Saxons, 177
Scientists' Center for Alternatives to Animal Tests, 61
Scoticorum Historiae (Boethius), 146
Scots, 177
Scott, Walter, 133–35, 148, 153n5; *Cadyow Castle,* 133–35, 139, 148; *The Minstrelsy of the Scottish Border,* 133
Scottish Farmer, 108
scrapie, 92, 95, 96
Sedgwick, Adam, 69–70, 181
Service, Robert, 147
Sewell, Anna, *Black Beauty,* 8
Shaw, George, 172
sheep, 15, 16, 36, 37, 39, 40, 41, 75, 92, 95, 103, 118, 133, 172, 173, 197–98, 208, *210,* 211; Dishley breed, 157, 159, 160, 161, 162, 171, 173, 191; foot-and-mouth disease and, 186–87; Jacob breed, 109; "Leicestershire Improved Breed," *167;* "mountain sheep" breeds, 191; origins of, in Britain, 193; shearing of, *165;* Soay, 193. *See also* Bakewell, Robert; Herdwick flock
Singer, Peter, 52–53, 54, 74, 88n3
slaughterhouse equipment, 60
sloth, 42
Society for Animal Rights, 85, 88n6

Society for the Prevention of Cruelty to Animals (SPCA), 54
Society for the Total Abolition and Utter Suppression of Vivisection, 77
South Africa, 91
South Korea, 91
Spinelli, Joseph, 84
Spira, Henry, 85
Sportsman, The, 107
squirrels, 38; domestication and, 41–42; in the Lake District, 196–97
Staffordshire, 138
"Standards for Research with Animals: Current Issues and Proposed Legislation" (conference), 73–74
Storer, John, 137, 153; *The Wild White Cattle of Great Britain*, 140
Stubbs, George, *A Comparative Anatomical Exposition of the Human Body with that of a Tiger and a Common Fowl*, 183–84; illustration from, *184*
Systema Naturae (Linnaeus), 179

Tankerville, 5th Earl of, 135, 139, 142, 145–46, 147, 153n7
Tankerville, 6th Earl of, 135, 136, 137, 148
Tankerville, 7th Earl of, 153
telegony, 21, 110, 116, 117, 118, 120, 128
Teltruth, T., 31, 39
tigers, 41; danger of, 43–44, *45;* hybrid lion-tiger, 114, *119*
Times of London, 104, 108
Topsell, Edward, *Historie of Foure-Footed Beastes*, 30, 47n10
Transactions of the Tyneside Naturalists Field Club, 149
Trimmer, Mary, 38
Trimmer, Sarah Kirby, 35, 38; *Fabulous Histories*, 38

Trull, Frankie, 85
tsetse flies, 108–9, 112
Tufts School of Veterinary Medicine, 84, 86
Tyson, Edward, 7, 65, 69, 178; *Orang-outang*, 65, 178, 179; —, frontispiece, 66

unicorn, 33
Uniformed Services University of Health Sciences, 87
United States, 92, 98–99, 101, 207, 209
University of California, San Francisco, 84
University of Edinburgh, 106, 109, 146

Variation of Animals and Plants under Domestication, The (Darwin): audience for, 126–27, 131; background of, 124; breeding and, 128–29; chicken in, 128; Chillingham cattle and, 147; cattle and, 127; cats and, 127; connection to *The Origin of Species*, 126; description and arguments of, 126–27, 128, 129–30, 131; dogs and, 126–27; domesticated plants and, 127; genetics in, 129–30; hybrids and, 129; *The Origin of Species* and, 129; and pangenesis theory, 130; pigeons and, 127; problems with, 128, 130; title page of, *125;* as Victorian era document, 130–31
Varty, Thomas, *Graphic Illustrations of Animals*, 36
Vasey, George, 141, 149; *Monograph of the Genus Bos*, 141
Veterinarian, 113
Victoria, Queen, 75, 104
Vindication of the Rights of Women (Wollstonecraft), 10
vivisection, 50, 54–55, 70, 73, 84,

108–9; arguments over, 50–52; as compared to antivivisection movement, 6; complexity of issue, 59–62; compromise with antivivisectionists, 61; Cruelty to Animals Act and, 77; Darwin and, 55, 57, 75; in the imagination, 56; litigation against, 51; media and, 87; medical advance and antivivisectionism, 83; modern debate over, 59–61, 83–88; Pasteur and, 81; protests against, 51, 61; popularity in United Kingdom, 55, 57; public demonstrations of, 53; RSPCA and, 75–76; scientists and, 51, 52, 55, 56, 57, 61–62, 74, 84–88; skepticism about, 57, 59. *See also* antivivisectionism

Wallace, Alfred Russel, 124
Wallace, R. Hedger, 145
Washington Post, 87
weasels, 33, 43
Wells, H. G., *The Island of Dr. Moreau*, 57–59
West Cumberland Fell Dales Sheep Association, 192
White Park Cattle Society, 141
wild animals, 9, 37, 75; brevity of information on, 41; British Empire and, 204; difference between domestic and wild animals, 200, 208–11; domestication and, 42, 209; man-eating and, 44; protection of, 206; *Variation of Animals and Plants under Domestication* and, 126–27. *See also* Chillingham cattle
Wild Cattle of Chillingham, The (Landseer), *136*, 136–37, 141
wildlife: British contribution to, 206; changing definitions of, 208–9, 210; Darwin and, 206; debates over, 206, 211; ecology and, 206–7; human encroachment on, 205, protection of, 206–7
Wild White Cattle of Great Britain, The (Storer), 140
Wilkinson, John, 173
Williams, Raymond, 207
Winners in Life's Race, The (Buckley), 35, 44
Wilson, Sir Jacob, 142
Wolf Hollow (Boston, Mass.), 203–4
Wollstonecraft, Mary, *Vindication of the Rights of Women*, 10
wolves, 33, 43, 114, 144, 180, 196, 203–5, 204, 211; attitudes toward, 204–5; protection of, 203–4, 207
Wordsworth, William, *Guide to the Lakes*, 195

Yellowstone National Park, 207, 208
Yonge, Sir Maurice, 109
York, 103, 104, 106, 114, 115, 118
Youatt, William, 13, 19, 190, 191, 192
Young, Arthur, 159

zebras, 37, 126, 172, 185, 209; Ewart's work with, 107, 111–13; *Guide to the Zebra Hybrids* (Ewart), 106, 114, 115, 120; —, illustration from, *113*; hybrids, 104, 107, 108, 110, 112–13, *113*, 119, 180; Matopo (zebra), 104, *105*, 107, 108, 111–12, 115; Romulus (hybrid zebra), 107
Zoological Journal, 68
Zoological Society of London. *See* Regent's Park Zoo
Zoologist, 113, 114
Zoophilist, The, 76, *76*
zoos, 7, 9, 33, 66, 109, 203, 209

— *Illustration Credits* —

Page 4, from the collections of the Ernst Mayr Library, Museum of Comparative Zoology, Harvard University; *pages 16, 31, 32, 35, 42, 45, 54,* Special Collections, University of Virginia Library; *page 66,* from the collections of the Ernst Mayr Library, Museum of Comparative Zoology, Harvard University; *page 67, In Memory of Consul,* n.p., reproduced with permission of Cheltham's Library, Manchester; *page 79,* Special Collections, University of Virginia Library; *page 111,* painting in the Hunterian Museum, Royal College of Surgeons, London, © The Royal College of Surgeons of England; *pages 125, 127,* Special Collections, University of Virginia Library; *page 136,* Laing Art Gallery, Tyne & Wear Archives and Museums; *page 138,* Special Collections, University of Virginia Library; *page 152,* Stephen Hall; *page 159,* Royal Agricultural Society of England; *page 165,* Bedford and Luton Archives and Records Service, Borough Hall, Bedford MK42 9AF; *page 167,* Special Collections, University of Virginia Library; *page 184,* Yale Center for British Art, Paul Mellon Collection; *page 204,* Special Collections, University of Virginia Library.